CMOS WIRELESS TRANSCEIVER DESIGN

THE KLUWER INTERNATIONAL SERIES IN ENGINEERING AND COMPUTER SCIENCE

ANALOG CIRCUITS AND SIGNAL PROCESSING

Consulting Editor
Mohammed Ismail
Ohio State University

Related Titles:

DESIGN OF LOW-VOLTAGE, LOW-POWER OPERATIONAL AMPLIFIER CELLS, *Ron Hogervorst, Johan H. Huijsing*, ISBN: 0-7923-9781-9

VLSI-COMPATIBLE IMPLEMENTATIONS FOR ARTIFICIAL NEURAL NETWORKS, *Sied Mehdi Fakhraie, Kenneth Carless Smith*, ISBN: 0-7923-9825-4

CHARACTERIZATION METHODS FOR SUBMICRON MOSFETs, edited by *Hisham Haddara*, ISBN: 0-7923-9695-2

LOW-VOLTAGE LOW-POWER ANALOG INTEGRATED CIRCUITS, edited by *Wouter Serdijn*, ISBN: 0-7923-9608-1

INTEGRATED VIDEO-FREQUENCY CONTINUOUS-TIME FILTERS: *High-Performance Realizations in BiCMOS*, *Scott D. Willingham, Ken Martin*, ISBN: 0-7923-9595-6

FEED-FORWARD NEURAL NETWORKS: *Vector Decomposition Analysis, Modelling and Analog Implementation*, *Anne-Johan Annema*, ISBN: 0-7923-9567-0

FREQUENCY COMPENSATION TECHNIQUES LOW-POWER OPERATIONAL AMPLIFIERS, *Ruud Easchauzier, Johan Huijsing*, ISBN: 0-7923-9565-4

ANALOG SIGNAL GENERATION FOR BIST OF MIXED-SIGNAL INTEGRATED CIRCUITS, *Gordon W. Roberts, Albert K. Lu*, ISBN: 0-7923-9564-6

INTEGRATED FIBER-OPTIC RECEIVERS, *Aaron Buchwald, Kenneth W. Martin*, ISBN: 0-7923-9549-2

MODELING WITH AN ANALOG HARDWARE DESCRIPTION LANGUAGE, *H. Alan Mantooth, Mike Fiegenbaum*, ISBN: 0-7923-9516-6

LOW-VOLTAGE CMOS OPERATIONAL AMPLIFIERS: *Theory, Design and Implementation*, *Satoshi Sakurai, Mohammed Ismail*, ISBN: 0-7923-9507-7

ANALYSIS AND SYNTHESIS OF MOS TRANSLINEAR CIRCUITS, *Remco J. Wiegerink*, ISBN: 0-7923-9390-2

COMPUTER-AIDED DESIGN OF ANALOG CIRCUITS AND SYSTEMS, *L. Richard Carley*, Ronald S. Gyurcsik, ISBN: 0-7923-9351-1

HIGH-PERFORMANCE CMOS CONTINUOUS-TIME FILTERS, *José Silva-Martínez, Michiel Steyaert, Willy Sansen*, ISBN: 0-7923-9339-2

SYMBOLIC ANALYSIS OF ANALOG CIRCUITS: *Techniques and Applications*, *Lawrence P. Huelsman, Georges G. E. Gielen*, ISBN: 0-7923-9324-4

DESIGN OF LOW-VOLTAGE BIPOLAR OPERATIONAL AMPLIFIERS, *M. Jeroen Fonderie, Johan H. Huijsing*, ISBN: 0-7923-9317-1

STATISTICAL MODELING FOR COMPUTER-AIDED DESIGN OF MOS VLSI CIRCUITS, *Christopher Michael, Mohammed Ismail*, ISBN: 0-7923-9299-X

SELECTIVE LINEAR-PHASE SWITCHED-CAPACITOR AND DIGITAL FILTERS, *Hussein Baher*, ISBN: 0-7923-9298-1

ANALOG CMOS FILTERS FOR VERY HIGH FREQUENCIES, *Bram Nauta*, ISBN: 0-7923-9272-8

ANALOG VLSI NEURAL NETWORKS, *Yoshiyasu Takefuji*, ISBN: 0-7923-9273-6

ANALOG VLSI IMPLEMENTATION OF NEURAL NETWORKS, *Carver A. Mead, Mohammed Ismail*, ISBN: 0-7923-9049-7

AN INTRODUCTION TO ANALOG VLSI DESIGN AUTOMATION, *Mohammed Ismail, José Franca*, ISBN: 0-7923-9071-7

CMOS WIRELESS TRANSCEIVER DESIGN

by

JAN CROLS
Katholieke Universiteit Leuven,
Heverlee, Belgium

and

MICHIEL STEYAERT
Katholieke Universiteit Leuven,
Heverlee, Belgium

KLUWER ACADEMIC PUBLISHERS
BOSTON / DORDRECHT / LONDON

A C.I.P. Catalogue record for this book is available from the Library of Congress.

ISBN 0-7923-9960-9

Published by Kluwer Academic Publishers,
P.O. Box 17, 3300 AA Dordrecht, The Netherlands.

Sold and distributed in the U.S.A. and Canada
by Kluwer Academic Publishers,
101 Philip Drive, Norwell, MA 02061, U.S.A.

In all other countries, sold and distributed
by Kluwer Academic Publishers,
P.O. Box 322, 3300 AH Dordrecht, The Netherlands.

Printed on acid-free paper

Printed in the Netherlands

Contents

Symbols, Conventions, Notations and Abbrevations

Conventions and Notations

The following notations are used for the subscripts of voltage and current signals to indicate their instantaneous, AC or DC value. The notation method of [Laker 1994] is used.

I_{OUT}	DC or average value of a current signal;
I_{out}	amplitude of the AC-component of a current signal in steady state;
i_{out}	instantaneous value of the AC component of a current signal;
i_{OUT}	total instantaneous value of a current signal, so $i_{OUT} = I_{OUT} + i_{out}$.

When the unit `dBm` is used throughout this text, it is not used in its original definition of 0 dBm being equal to 1 mW in 50 Ω. Unlike in discrete realizations, integrated RF systems often use impedance levels that differ from 50 Ω. In order to allow a comparison with classical discrete RF design, the unit `dBm` is still used, however redefined as the corresponding voltage level in 50 Ω systems. 0 dBm is thus defined as 223 mV_{rms} independent of the impedance level. 20 dBm is equal to 2.23 V_{rms}.

Bibliografic References

In this text, bibliographic references contain information on the first author, the publication source and the year of publication, possibly extended with an extra character when more than one publication of the same author has been published in the journal in the same year. In this way the reader finds already a lot of information on bibliographic references within the text. The full information can of course be found in the bibliography. An example is [Crols JSSC95a]. 'Crols' are the first five letters of the first author's name. 'JSSC' is an abbrevation for the journal the reference was published in. '95' is the year of publication and the character 'a' has been added to avoid ambiguity with another bibliographic reference. A list of the most important abbrevations used for publication sources has been given below. The absense of such an abbrevation in a reference indicates that it refers to a book.

ACD	Analog Circuit Design
CASII	IEEE Transactions on Circuits and Systems II
CICC	proceedings of the Custom Integrated Circuits Conference
ESSC	proceedings of the European Conference on Solid-State Circuits
ISSCC	digest of technical papers of the International Conference on Solid-State Circuits
JSSC	IEEE Journal of Solid-State Circuits
VLSI	proceedings of the Symposium on VLSI Circuits

Symbols

The symbol convention given below is used in the circuit schematics throughout this text. Unless otherwise indicated, the bulk of nMOS transistors is always assumed to be connected to the ground and the bulk of pMOS transistors is always assumed to be connected to their source.

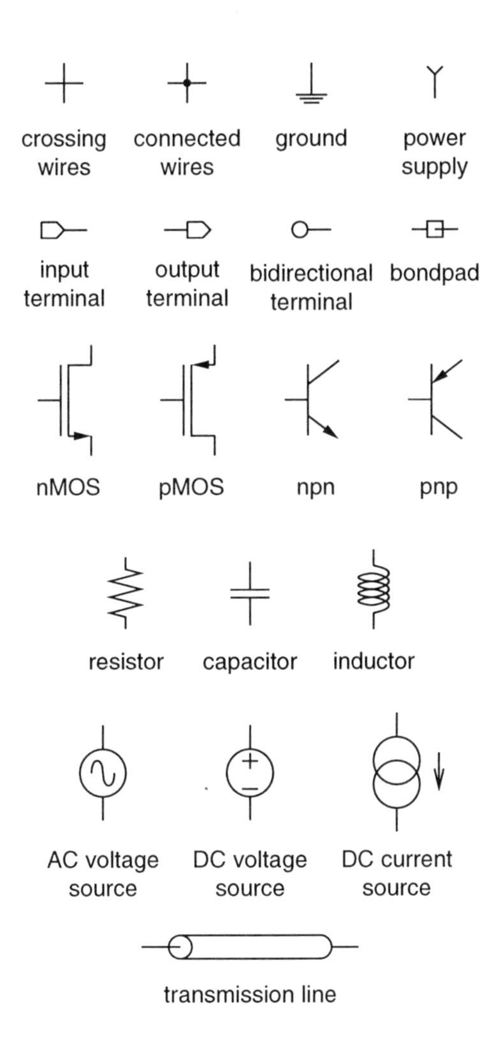

Abbreviations

This list gives the full description of the abbreviations used troughout the text.

ADC	analog-to-digital converter
AC	alternating current
AGC	automatic gain control
BB	baseband
BER	bit error rate
BPF	bandpass filter
CMOS	complementary MOS
CPU	central processor unit
DAC	digital-to-analog converter
DC	direct current
DDS	direct digital synthesis
DECT	digital European cordless telephone
DR	dynamic range
DSP	digital signal processor
EEPROM	electrically erasable programable ROM
FER	frame error rate
FFT	fast Fourier transform
FM	frequency modulation
FS	frequency shift
GaAs	gallium arsenide
GMSK	gaussian minimum shift keying
GSM	global system mobile
HDn	n^{th} order distortion
HF	high frequency
I	in phase
IF	intermediate frequency
IMn	n^{th} order intermodulation
IPn	n^{th} order intermodulation intersection point
ISDN	integrated services digital network
LF	low frequency
LNA	low noise amplifier
LO	local oscillator
LPF	lowpass filter
LTF	linear transfer function
modem	modulator-demodulator
MOS	metal oxide semiconductor
MOSFET	MOS field effect transistor
NF	noise figure

OTA	operational transconductance amplifier
PLL	phase locked loop
PSD	power spectral density
Q	in quadrature
QPSK	quad phase shift keying
RBER	residual bit error rate
RF	radio frequency
RMS	root mean square
ROM	read only memory
$\Sigma\Delta$	sigma-delta
SNR	signal-to-noise ratio
spec	specification
SSB	single sideband
SUSR	signal-to-unwanted-signal ratio
TDMA	time domain multiple access
transceiver	transmitter-receiver
VCO	voltage controlled oscillator
VGA	variable gain amplifier
Xtal	crystal

Preface

The world of wireless communications is changing very rapidly since a few years. The introduction of digital data communication in combination with digital signal processing has created the foundation for the development of many new wireless applications. High-quality digital wireless networks for voice communication with global and local coverage, like the GSM and DECT system, are only faint and early examples of the wide variety of wireless applications that will become available in the remainder of this decade.

The new evolutions in wireless communications set new requirements for the transceivers (transmitter-receivers). Higher operating frequencies, a lower power consumption and a very high degree of integration, are new specifications which ask for design approaches quite different from the classical RF design techniques. The integratability and power consumption reduction of the digital part will further improve with the continued downscaling of technologies. This is however completely different for the analog transceiver front-end, the part which performs the interfacing between the antenna and the digital signal processing. The analog front-end's integratability and power consumption are closely related to the physical limitations of the transceiver topology and not so much to the scaling of the used technology. Chapter 2 gives a detailed study of the level of integration in current transceiver realization and analyzes their limitations.

In chapter 3 of this book the complex signal technique for the analysis and synthesis of multi-path receiver and transmitter topologies is introduced. With this technique, several new receiver and transmitter topologies are developed. An example is the low-IF receiver topology. The presented topologies all have in common that they combine the advantages of the classically used heterodyne architectures, i.e. a very good performance, with the advantage of a very good integratability.

Determining the building block specification for a new transceiver architecture is for RF designers mainly an experience-based process, resulting in long design cycles and only a very gradual advancement of transceiver architectures. Here, in chapter 4, a formal methodology for the high-level design of transceiver architectures is presented. This methodology allows a structured and computer automatic high-level design, resulting in short design cycles and a fast evaluation of new transceiver architectures. The full high-level design of a low-IF / direct upconversion GSM transceiver front-end is presented in chapter 5.

A true full integration of a wireless transceiver requires that the analog front-end is integrated on the same die as the transceiver's digital baseband signal processor. DSP's use however standard CMOS technologies and these are less performant than the silicon bipolar and GaAs technologies that are used today for the integration of analog transceiver front-ends. Therefore, the integration of RF building blocks in CMOS is studied in chapter 6. Several chip realization are presented. In chapter 7 the capabilities of deep sub-micron CMOS used in combination with new highly integrated transceiver topologies for the implementation of wireless transceiver front-ends in the 1 to 2 GHz range is studied and demonstrated.

We also wish to express our gratitude to all persons who have contributed to the realization of this book and to the research described in this book. We would like to thank Prof. W. Sansen and Prof. H. De Man for carefully proofreading the manuscript. We would like to thank J. Craninckx, P. Kinget, M. Borremans and J. Janssens for their contribution made to this research. Our thanks also goes to the IWT (The Flemish Institute for Research in Science and Technology) for funding of the research.

Finally, we thank our families for their support and patience. Without it this research and this book would not have been possible.

Jan Crols
Michiel Steyaert

Department of Electrical Engineering - ESAT MICAS
Katholieke Universiteit Leuven
Leuven, Belgium, 1997

1 WIRELESS COMMUNICATIONS

1.1 HISTORICAL OVERVIEW

A little more than 100 years ago, in 1894, Guglielmo Marconi began the research that would lead to the development of the first practical wireless communication set-ups [EW92b]. Since then, the invention and development of wireless communication has had an enormous influence on the cultural and social evolution of 20th century society. Radio and television have become mass products and media with an impact on everybody's daily life that should not be underestimated.

In its early days, wireless communication was first used for maritime applications. Large ships were at that time the main long distance transportation method and also the only moving objects which were capable of carrying the large equipment and the enormous antenna for what we call today 'low frequency' radio communication. Since then, wireless communication equipment has gone through an enormous evolution. The introduction of new components, like the transistor, and the use of ever higher operating frequencies have led in the seventies to the development of small handheld portable radio equipment which is even today still widely used by e.g. emergency workers, like police officers, fire men and paramedics. Fig. 1.1a shows such a portable radio.

Despite the enormous evolution wireless communication has gone through and despite its widespread applications, it is important to remark that true bi-directional wireless communication, where each point has both a receiver and transmitter, has always been limited to professional applications. Today, wireless information services for a broad public (like radio and television programs) are only a uni-directional system in which a few stations provide information that can only be consumed in a passive way, i.e. without any means of interaction. Non-professional bi-directional communication has been limited to the small circle of highly dedicated amateur radio hobbyists. The general public uses wireline telephony for bi-directional communication.

The reason why bi-directional wireless communication has been limited to professional applications is because of the fragile nature of the communication medium : free air. It has to be shared by every application and every user within the application. Any evil-minded or unknowledgable person can easily disturb a wireless communication. Those who are going to use a wireless application therefore have to follow a strict communication code (based on e.g. calling codes and call signs). Often one has to take training and pass an exam first before one is allowed to transmit anything. This has limited bi-directional communication to mainly professional use only. The importance of the influence of uni-directional public radio and television broadcasting on the 20th century can not be denied. One can wonder however what the influence could have been when bi-directional interactive wireless communication had been accessible for the large mass.

1.2 WIRELESS NETWORKS

Today, the introduction of digital signal processing techniques is bringing a lot of changes to the world of wireless communications. Digital signal processing in combination with digital data transmission allows for the use of highly sophisticated modulation techniques, complex demodulation algorithms, error detection and correction techniques and data encryption, resulting all together in a huge improvement of the quality of the wireless link. Even more important however is the fact that digital communication also allows for the use of digital data flow control. Microprocessor-based systems can be used to transmit and receive data in frames where each frame is labelled with an identification tag and routed through a virtual wireless network.

The principle of the wireless network is a very important aspect of modern wireless applications. A wireless network is a system in which there is an automated point-to-point communication between two or more users possible without interference of the other users. Although all users are using the same communication medium, a transparent communication service is set up in which it appears to each user that he can have a direct link with any other user. The importance of the possibility to set up wireless networks may not be underestimated. It makes the need for a strict communication protocol between users unnecessary and in this way it makes bi-directional wireless

(a) (b)

Figure 1.1. Two examples of handheld wireless equipment. a) a classic portable radio for professional use and b) the new type portable cellular phone for public use. The new cellular phones use digital data transmission and they have an automated access to the wireless network.

communication for the general public possible. Examples of wireless network are cellular networks for voice telephony like GSM. Fig. 1.1b shows a photo of a portable handset for wireless telephony. The introduction of the GSM system has already had in many countries a serious impact on the personal communication of people and yet it is only a faint example of the many new wireless services for personal use that may be expected in the future. Applications will range from low-cost wireless e-mail services, over wireless identification and tracking systems (RFID), to full-blown portable multi-media terminals. These services will not only be available for a selected group of users, they will have a very wide penetration and this will in turn also have a large impact on the information content that will be available via these services. The predicted market growths are enormous [Neuvo ESSC96].

Transparent wireless networking is only one aspect necessary for a widespread introduction of wireless communication services. Cost price is of course also a very important aspect [Wenin ESSC94]. When wireless terminals can not be introduced at a price below 300 US$, their use will still remain limited to privileged users. The price of wireless interfaces that are integrated as part of a much larger system, like for instance a portable computer or personal digital assistant, will have to be even lower. Prices of 100 US$ and lower per wireless interface must be achieved. Another important aspect is weight (or volume) and linked to this power consumption (a lower power consumption means smaller, and thus cheaper and lighter, batteries). Wireless terminals are only interesting for a large public when they can be used practically anywhere and at any time. This implies handheld terminals with a low weight (below 200 g). Again, terminals part of a larger system will have to have an even lower weight (down to 50 g and below).

1.3 FULL INTEGRATION AND CMOS TRANSCEIVERS

These new evolutions in wireless communications set new requirements for the wireless interfaces, the transceivers (transmitter-receivers). A very high degree of integration, a lower power consumption and the use of a lower power supply are the goals set for new developments in wireless transceiver design. The integratability and power consumption reduction of the digital part will further improve with the continued downscaling of technologies. It is also the digital part that benefits most from a reduction of the power supply voltage. The bottleneck for further advancement is the analog front-end. The analog front-end forms the interface between the antenna and the digital signal processor . For the analog front-end, integratability and power consumption reduction are closely related to the physical limitations of the transceiver topology and to the used technology.

Present-day transceiver often consist of a three or four chip-set solution combined with several external components. This is evolving rapidly to a two chip-set solution (one for the analog signal processing and one for the digital signal processing) with some external components [Stetz ISSCC95, Marsh ISSCC95]. The required number of external components is linked to the physical limitations of the analog front-end topology. A further reduction of the number of external components is essential to obtain a lower cost, power consumption and weight, but it will require a fundamental change in the design of analog front-end architectures. The development of new front-end architectures that can be fully integrated (without requiring any external component) and which have at the same time a high performance, is necessary for the further evolution to a new generation of wireless transceivers. In this book this topic is researched and several new architectures are introduced.

An even further step in the evolution of wireless transceiver design will be the realization of the complete transceiver (analog and digital signal processing) on one chip. Apart from the problem of interference of digital switching noise on the analog signal processing, which is a problem for the realization of any mixed analog-digital design, there is also a problem of incompatible technologies. The analog front-end requires a high performance technology, like GaAs or silicon bipolar, with devices which can easily achieve operating frequencies of more than 2 GHz. For the digital signal processor a small device feature size is essential. It implements complex algorithms and this requires the implementation of a huge number of transistors. Therefore, only the best submicron CMOS technologies allow for a feasible and cost-effective integration of the DSP. The use of a BiCMOS technology for the integration of a complete wireless transceiver is not a solution. The chip area occupied by the analog part is only a fraction of the chip area occupied by the digital part. The feature size of the MOS transistors is even in the best currently available BiCMOS processes always a bit larger than in the state-of-the-art CMOS processes. Using a BiCMOS process to improve the performance of a very small part of the transceiver (the analog front-end) will go at the cost of a serious increase of the total chip area, which will result in an unacceptable cost increase. The evolution of CMOS technologies to sub-micron and deep sub-micron can bring a solution to this problem. Achieving frequencies of more than 2 GHz becomes also possible with the devices of these new advanced technologies. The use of CMOS technologies for the realization of analog transceiver front-ends will require the development of new circuit design techniques in order to achieve a transceiver performance equal or better than present-day realizations. In this book circuit design techniques for high frequency analog transceiver design in CMOS technologies are researched and discussed. Many chip realizations are presented.

1.4 THE PRESENTED WORK

The goal of the research project described in this book is to realize a fully integrated one-chip transceiver to be used in portable and low cost applications. The emphasis lies on achieving a very high degree of integration (a true single-chip solution which does not require external components, nor any tuning or trimming) without doing this at the cost of a lower performance (target applications are GSM, DCS 1800, DECT, ...) or higher power consumption. The second goal is the realization of this transceiver in a CMOS technology, a vital step towards a low cost implementation and the future integration of the complete transceiver system (both the analog and digital part) on one chip. In this book results on the following research topics are therefore presented :

- **The development of new receiver topologies**
  Both the classically used IF and zero-IF receiver topologies have their advantages and disadvantages concerning integratability and performance. Therefore, a new receiver topology, called the low-IF receiver, is developed as part of this research. Not only is this new receiver topology developed, but at the same time a new analysis and synthesis technique for receiver architectures is proposed. It is based on the use of complex signals and gives a clear insight in all the effects that reduce the performance of a receiver. With this technique several new receiver topologies are synthesized. These new topologies all have the property that they combine the advantages of the two classical receiver structures : a very high degree of integration with a high performance, without an increase of the power consumption.

- **Characterization of the relationship between the performance and the power consumption of a receiver**
  The minimization of the power consumption reduces also the battery size and cost

in portable applications. A full characterization of all sources of receiver performance degradation (noise, distortion, crosstalk, ...) is done and a model for their relationship with the power consumption is worked out. An important aspect of the presented research is the development of high-level behavioral models for the different types of receiver building blocks which describe the relationship between their high-level specifications and their power consumption.

- **Development of a high-level synthesis technique for receiver topologies**
  With the aid of high-level behavioral models for building blocks, it becomes possible to develop an optimization method for receiver topologies. This makes a fast evaluation of the qualities of a certain receiver topology for a certain application possible and it enables the automated translation of high-level receiver specification to the most optimal building block specifications.

- **The development of new transmitter topologies**
  The degree to which transmitters are integrated today, is still fairly low. Critical parts are the upconverter and the power amplifier. In this book new transmitter topologies are developed and presented. The main goals are a low sensitivity to non-linearities and an excellent power efficiency at high frequencies.

- **The design, realization and testing of new types of building blocks needed for the newly developed receiver and transmitter topologies**
  The main part of the presented research is the actual design and chip realization of building blocks for transceivers. With these chip realizations the feasibility of new building block types, which are necessary for the development of the new receiver and transmitter topologies, is illustrated.

- **The design, realization and testing of high performance CMOS circuits for high frequency transceivers**
  A second aspect to the presented chip realizations is that they are always implemented in full CMOS processes. In this way the feasibility of the use of CMOS processes for high frequency (1 to 2 GHz) RF applications is illustrated. Many new techniques have to be developed to achieve these high-frequencies with today's sub-micron processes. The goal is not only the realization in CMOS, but also the achievement in CMOS of performances which are better than the existing alternatives.

- **The realization of a fully integrated one-chip transceiver**
  The final goal of the presented research is a first realization of a fully integrated one-chip transceiver in CMOS which requires no external components. This realization can be used to study and evaluate the interaction between its different building blocks and in this way it can form a starting point for future research on

the realization of a fully integrated transceiver. In order to achieve a high performance, this chip uses the presented newly developed topologies. The specifications for the different building blocks are such that the overall power and area consumption is minimal. As an example and in order to show that the developed methods can indeed be used for high performance applications, the chip is designed to fully comply with the GSM specifications.

2 TRANSMITTERS AND RECEIVERS

2.1 INTRODUCTION

A few years ago the world of wireless communications and its applications started to grow rapidly. The introduction of digital data transmission in combination with digital signal processing and a digital signal flow control, has created the breeding ground for the development of many new wireless applications. High quality digital networks for voice communication with global and local coverage, like the GSM and the DECT system, are only faint and early examples of the wide variety of wireless services that will become available on a large scale in the remainder of this decade. Applications will vary from very cheap low-end wireless e-mail services, over general purpose wireless identification and tracking systems, to fully equipped high performance portable multi-media terminals.

These new evolutions in wireless communications set new requirements for the transceivers used in its applications. Higher operating frequencies, a lower power supply voltage, a lower power consumption and a very high degree of integration, are new design specifications which ask for design approaches quite different from the classical RF design techniques. New digital techniques allow for the use of complex modulation schemes and elaborate demodulation, encryption and error correction al-

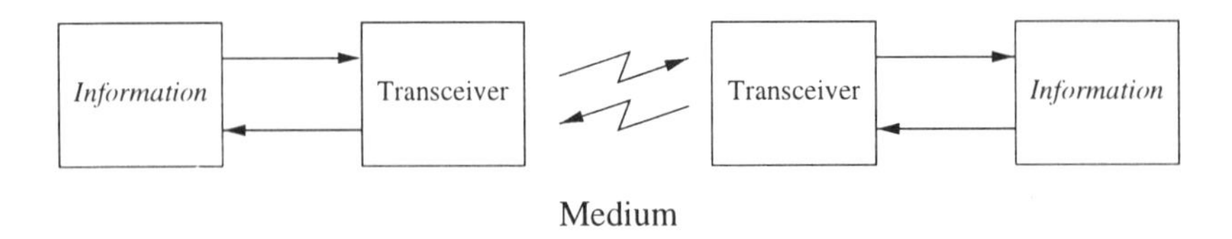

Figure 2.1. The basic configuration of a communication system, showing transceivers and the communication channel.

gorithms, resulting in high-quality wireless links with low bit-error rates. In these new transceivers it is the analog transceiver front-end, the part which performs the interfacing between the antenna and the digital signal processing, that forms the bottleneck for further advancement. The analog receiver has to convert in successive stages of filtering, amplifying and downconverting, a high frequency, very high dynamic range signal into a low frequency, low dynamic range signal that can be sampled easily. The transmitter has to upconvert a signal to the same high frequency with a very good linearity and a large output power. Integratability and power consumption reduction of the digital part will further improve with the continued downscaling of technologies, but this is completely different for the analog front-end, for which integratability and power consumption are closely related to the physical limitations imposed by the transceiver topology and not by the used technology.

2.2 TRANSCEIVERS

A transceiver (transmitter-receiver) is the two-way interface between an information source and the communication channel which will be used to exchange the information. Fig. 2.1 shows this basic configuration that holds for any communication system. The transmitter translates a data stream into a form which is suited for the communication channel. This translation is called modulation. During the process of modulation, information will be reordered and reformatted. It is also possible that extra information is added. In this way the signal can be adapted to the specific needs of the communication channel concerning frequency and time domain spectrum. A good adaptation of the transmitted signal to the channel is necessary to obtain a high quality communication link. The receiver performs the reverse operation, called demodulation.

There are many different types of communication systems. Examples vary from twisted-pair analog telephony, over optical fiber broadband ISDN networks, to cell-based digital wireless mobile networks. Each system has its typical behavior and therefore has its own demands for such things as frequency spectrum, time spectrum, data parallelism and required communication quality. All wireless applications have two aspects in common : they operate in a passband around a carrier frequency and the carrier frequencies are constantly increasing with every newly introduced application (GSM operates for instance in a 30 MHz band around 890 MHz). Both are closely related to the fact that all wireless applications have to share their medium : electromagnetic waves through open air. Band limited operation prevents interference between the different applications and the use of higher and higher operating frequencies is, in combination with lower transmit radii and smaller cells, the only way to obtain more bandwidth. Both aspects have a very big impact on the specific nature of transceivers for wireless applications.

2.3 INTEGRATED RECEIVERS

The specific nature of wireless communication channels (passbands around high operating frequencies) makes that almost all receiver types for wireless applications consist of two main parts : a front-end which performs the frequency downconversion and a back-end part in which the actual demodulation of the signal is performed. A third part, called the user-end, transforms the received information into a form suitable for the user. This part is however not considered to be part of the receiver. Over the years many different receiver types have been developed for many different applications. Their basic way of operation is however always the same. Fig. 2.2 shows this basic configuration : the downconverter brings the antenna signal down from its high operating frequency to a suitable, lower, frequency on which then the demodulation can be performed in an elegant way (remark that the values indicated on this figure and on any other receiver or transmitter block schematic shwon in this chapter are not directly related to an existing realization; they are shown to give the reader an indication of their magnitude in a typical application). The antenna signal consists of a very broad spectrum of many different information channels and noise sources. The wanted signal is, in its modulated form, only a very small part of this broad spectrum. The front-end part converts this antenna signal in a signal that can be demodulated by the back-end part in a feasible way.

2.3.1 Downconversion in the Front-End

The receiver front-end has to be able to downconvert very small signals (e.g. a bandwidth of a few 100 kHz and an amplitude which can be as low as 5 μV_{rms}) from a high operating frequency (typically between 900 MHz to 2.4 GHz) to a much lower intermediate frequency (typically 10 MHz or lower) or to the baseband, where it can be demodulated (more information on the specifications for wireless communication system is given in chapter 5). This small signal is present in a noisy environment and it can be surrounded with other signals which can be as large as 300 mV_{rms}. This requires from the front-end a high operating frequency and, at the same time, a high input dynamic range. These specifications make that the front-end is always realized as an analog circuit which mainly consists of mixers, filters and variable gain amplifiers which, in successive stages, further and further filter the antenna signal and downcon-

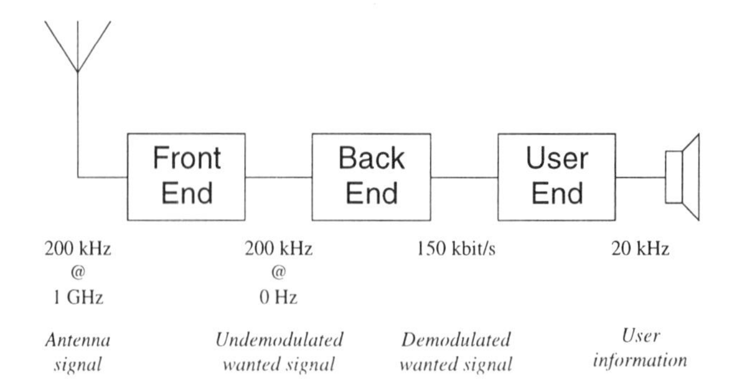

Figure 2.2. The basic configuration of a receiver for wireless communications.

vert it to lower frequencies to become finally a low frequency signal that consists only of the modulated wanted signal.

2.3.2 Demodulation in the Back-End

The demodulation technique that will be used in the back-end part of the receiver will mainly depend on two things : the modulation technique that is applied and the type of front-end that is used. It is obvious that the type of demodulator depends on the modulation that has been performed : for an FM modulated signal an FM demodulator is required and for a QPSK modulated signal a QPSK demodulator is required [Shanm 1979, Grone TCOM76, Muro TCOM81].

A front-end has no influence on the form of the wanted signal (see section 2.3.1), but it does determine the center frequency at which the wanted signal is provided to the back-end part. There are two possibilities. The wanted signal is provided either on an IF frequency or it is provided as a quadrature baseband signal.

Today, most new wireless applications use digital data communication. Digital wireless has many advantages such as high quality communication due to signal regeneration and error correction systems, computer control of the communication and integration of different services in one application. Fig. 2.3 shows the block schematic of a typical transceiver for a digital wireless application [Stetz ISSCC95, Marsh ISSCC95]. In a receiver for a digital wireless application the demodulation is always performed in a digital signal processor (DSP) [Salla ISSCC90]. Almost all these DSP's perform the demodulation on a quadrature baseband signal. The front-end for a digital wireless receiver must therefore produce these baseband quadrature I and Q signals. This however not necessarily has to be done in an analog way. The wanted signal may be downconverted to an intermediate frequency (IF, e.g. 10.7 MHz). After sampling, it can then be further downconverted to the baseband in the DSP [Chien JSSC94]. Thus, in this case, the final part of the front-end is digital. This situation is shown in fig. 2.4.

2.4 RECEIVER FRONT-END ARCHITECTURES

2.4.1 Heterodyne or IF Receivers

The heterodyne receiver is without any doubt the most often used receiver topology [Shanm 1979, Okan TCE82]. It has been in use for a long time already and its way of operating is very well known. Its main feature is the use of an intermediate frequency (IF frequency). For this reason the heterodyne receiver is often also called the IF receiver. The latter phrasing will be used in this text.

2.4.1.1 The Single-Stage IF Receiver. In fig. 2.5 the schematic representation of a single-stage IF receiver's topology is given. The wanted signal is downconverted from its carrier frequency to the intermediate frequency by multiplying it with a single sinusoidal signal. It can then be demodulated on this frequency or it can be further downconverted. In a single stage version it will not be further downconverted.

The main disadvantage of the IF receiver is that, apart from the wanted signal, also an unwanted signal, called the mirror frequency, is converted to the IF. This is illustrated with fig. 2.6. When the wanted signal is situated on f_c, the mirror frequency is at $f_c + 2f_i$. Equation 2.0 gives the calculation of this effect. The mixer gives a frequency component on f_i for $f_x = f_c$ and $f_x = f_c + 2f_i$. The antenna signal $a(t)$ is in this calculation a phase modulated signal and equal to $\cos(2\pi f_x t + m(t) + \phi)$.

$$\begin{aligned} & a(t) \times \sin(2\pi(f_c + f_i)t + \psi) \\ = & \cos(2\pi f_x t + m(t) + \phi) \times \sin(2\pi(f_c + f_i)t + \psi) \\ = & \tfrac{1}{2}[\sin(2\pi(f_x + f_c + f_i)t + m(t) + \phi + \psi) \\ & - \sin(2\pi(f_x - f_c - f_i)t + m(t) + \phi - \psi)] \end{aligned} \tag{2.1}$$

The mirror frequency has to be suppressed before it is mixed down to the IF. This is done by means of a high frequency (HF) filter. Such an HF filter can only be realized when the intermediate frequency f_i is high enough because the wanted signal (on f_c) must be relatively far away from the mirror frequency. Even when the ratio f_c/f_i is as small as 10, the specifications for the HF filter are very severe. The HF filter must have a very high quality factor Q (50 to 100, Q is defined as the center frequency over the bandwidth), it must have a sufficiently high order (up to 6th order for high quality applications) and in some cases the center frequency must be tunable. A filter with these specifications can not be integrated. The accuracy requirements would, for a Q of 20 or more, be much too high for an integrated filter because their power consumption is proportional to Q^2 [Groen CAS91, Voorm 1993], resulting in totally unacceptable power consumption specifications. HF filters are therefore always real-

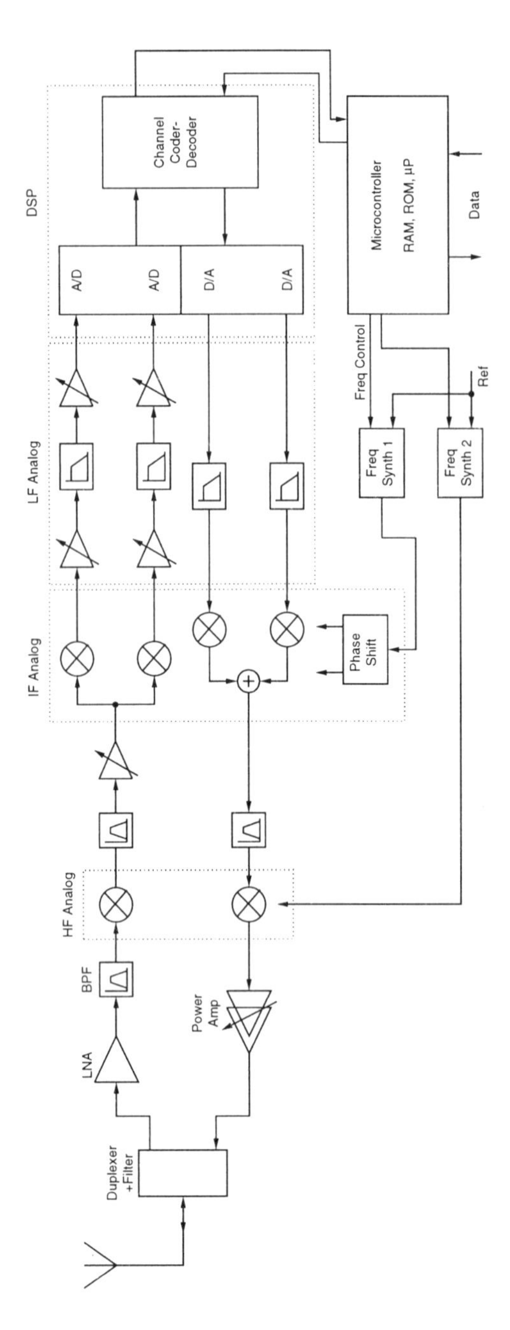

Figure 2.3. The typical transceiver topology for digital wireless communications.

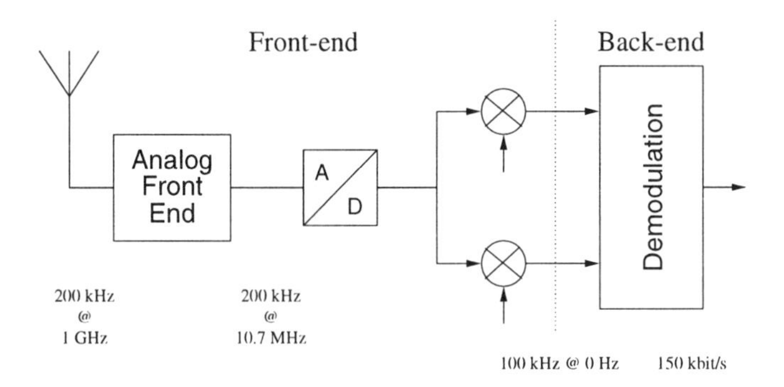

Figure 2.4. Front-end where the final downconversion from IF to baseband is performed in the digital domain.

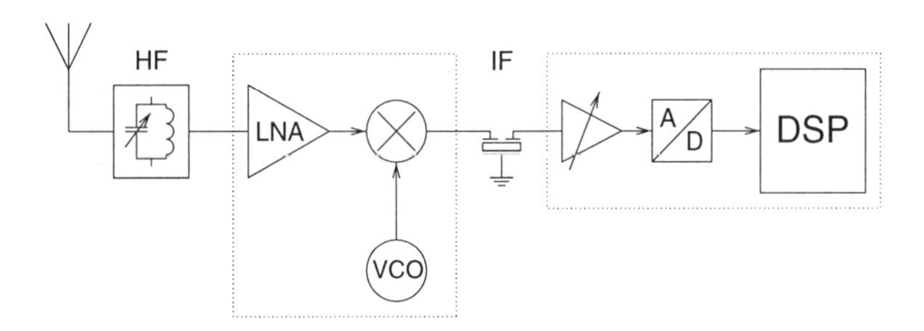

Figure 2.5. Block diagram of a single-stage IF receiver.

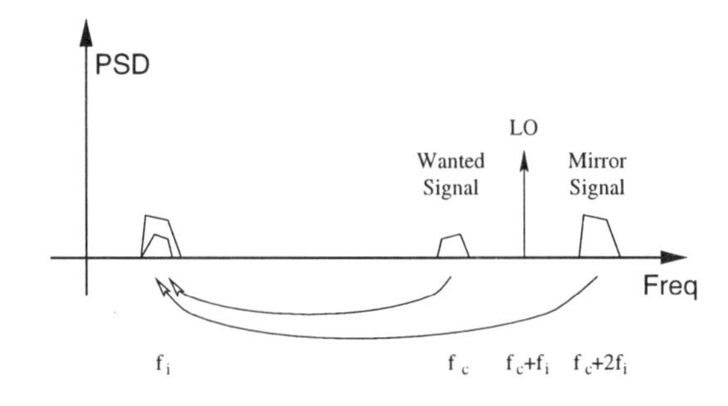

Figure 2.6. The frequency translations performed by mixing in a single-stage IF receiver.

ized as discrete off-chip components, but these are very expensive and vulnerable in use. Off-chip components require extra handling, extra board space and an increased pin-count and they reduce board yield. Their power consumption is high because they have to be driven at a low impedance (e.g. 50 Ω) to compensate for the large parasitic capacitances inherently present due to their large physical size.

Once the signal is downconverted to the IF, it has to be further filtered in order to obtain only the wanted signal. This filter must also have a sufficiently high Q (e.g. 50) and order (8th or 10th order). Integrating these IF filters is also very hard. Although ever more analog integrated IF filters are published [Silva JSSC92, Silva KUL92, Wang ISSCC89, Wang JSSC90], due to their intrinsic lower power consumption ceramic resonators are still used in most applications. Again, these are high quality discrete components and their use is very expensive compared to integrated components.

2.4.1.2 Multi-Stage IF Receivers. The operating frequency of newly introduced wireless applications is constantly increasing (from 900 MHz, to 1.8 GHz, to 2.4 GHz, to ...). The signal bandwidths however stay the same (between 200 kHz and 2 MHz). The same goes for the total bandwidth reserved for the application (10 to 50 MHz). The ratio between the operating frequency and the wanted signal bandwidth increases and this makes it harder and harder to use a single stage IF receiver. Going for instance directly from a 900 MHz operating frequency to a 10 MHz IF is impossible. The ratio is about 100, which would require from the HF filter a quality factor Q of 1000, a value that can not even be realized with very accurate discrete components. The single stage IF receiver is therefore unsuited for all newly introduced wireless applications.

A solution is the use of more than one IF stage. From stage to stage the wanted signal is further and further downconverted until the final IF is reached. The ratio between the operating frequency before and after a downconversion is always kept lower than 10. In order to avoid the need for dynamic tuning of the HF filter it is most of the time even kept lower than 5. An example : a signal at 900 MHz may in a first stage be downconverted to a first IF of 250 MHz, from there it is further filtered and downconverted to 50 MHz and in a final stage it is then downconverted to 10 MHz. Each downconversion stage has the same mirror frequency problem typical for the IF receiver and extensive filtering between stages is still necessary. These filters have still a high center frequency and a high Q and they will therefore all be discrete. This is the major drawback of the multi-stage IF receiver : its signal needs to go through many off-chip components which makes it vulnerable, expensive and sensitive to external parasitic signals. Its power consumption will be increased due to the many off-chip components which have to be driven at a low impedance.

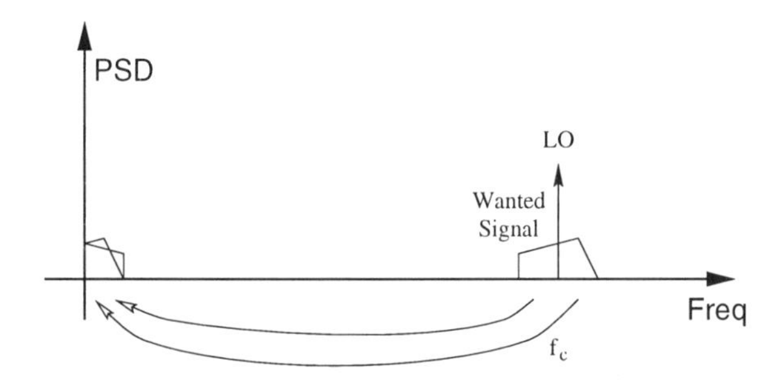

Figure 2.7. Frequency translations in an IF receiver when the IF is zero.

2.4.2 Homodyne or Zero-IF Receivers

In zero-IF receivers the wanted signal is directly downconverted to the baseband. The IF is chosen to be zero. In this case the signal on the mirror frequency is the wanted signal itself. This does however not eliminate the problem of the mirror frequency. Both signals, the wanted signal and the signal from the mirror frequency, are not the same. They are each others mirror image. Fig. 2.7 shows that this results in the lower and upper sideband being placed on top of each other in the baseband, which means that they become inseparable.

This problem is solved by doing the downconversion twice. Once with a sine and once with a cosine. This is called quadrature downconversion. The topology of a zero-IF receiver is given in fig. 2.8. In equations 2.2 and 2.3 the downconversion of an antenna signal $a(t)$ is calculated.

$$\begin{aligned} u_i(t) &= a(t) \times \cos(2\pi f_c t + \psi) \\ u_q(t) &= a(t) \times \sin(2\pi f_c t + \psi) \end{aligned} \tag{2.2}$$

The original signal $m(t)$ can, in case of phase modulation (see equation 2.0), be found as the angle of the vector $(v_i(t), v_q(t))$, a lowpass filtered version of $(u_i(t), u_q(t))$.

$$m(t) = \arctan\left(\frac{v_q(t)}{v_i(t)}\right) + \psi - \phi \tag{2.3}$$

The demodulation in a DSP can be done by means of an angle measurement algorithm. The CORDIC algorithm is an example of such an algorithm [Havi JSSC80].

The advantages of zero-IF receivers are obvious. There is no need for high Q HF or IF bandpass filters to suppress mirror signals. Nevertheless, in most zero-IF receiver designs an HF filter is still used to reduce the dynamic range requirements for the downconverter and prevent mixing of the RF signal with harmonic components of the LO signal. The lowpass filters, required after the downconversion mixers, can easily be

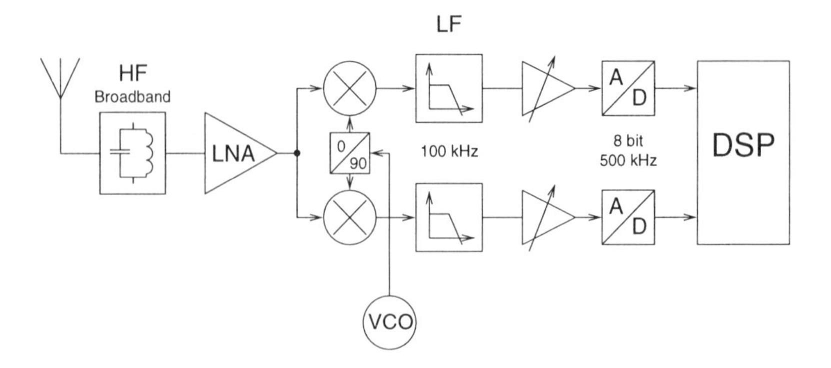

Figure 2.8. Block diagram of the zero-IF receiver.

realized as analog integrated filters. Another advantage of the zero-IF receiver is that the downconverted signals are delivered to the DSP as baseband I and Q quadrature signals, which results in the lowest possible A/D-converter requirements.

The precision with which both demodulation paths (I and Q) can be matched determines how good the mirrored signal can be suppressed. The specifications on mirror suppression for a zero-IF receiver are not as severe as they are for an IF receiver. In an IF receiver extra suppression is needed because on the mirror frequency the signal can be bigger than the wanted signal. This is not possible in the zero-IF receiver where the mirror signal is a frequency mirrored version of the wanted signal. However, a mirror signal suppression of 25 dB is still needed in a good quality zero-IF receiver [Rab ACD93]. The zero-IF receiver is thus sensitive to matching and to phase and amplitude error in the quadrature oscillator. It is due to this need for good matching properties that the implementation of the zero-IF receiver has only become feasible since the introduction of the trend towards on-chip integrated design of receivers.

Another problem in zero-IF receivers are parasitic baseband signals which are created during downconversion. They are mainly the result of the crosstalk between the RF and LO inputs of the mixers. The multiplication of the LO signal with itself results in a DC (or almost DC) signal. This DC-offset is superimposed on the wanted signal in the baseband. It can only be removed by means of very long time constants (at least 1/10 of a second [Brown ICLMR85]) and it always results in the loss of a part of the wanted signal. This has an effect that is comparable to distortion because low frequency information correlated to the wanted signal is removed [Brown ICLMR85]. The distortion will be lower and of an acceptable level if the time constant is longer (e.g. 1 s), but this long time constant makes the settling time of the complete receiver system too long. For instance, at each change of the carrier frequency the receiver would have to settle for at least a second when a highpass filter of 1 Hz is used. Fur-

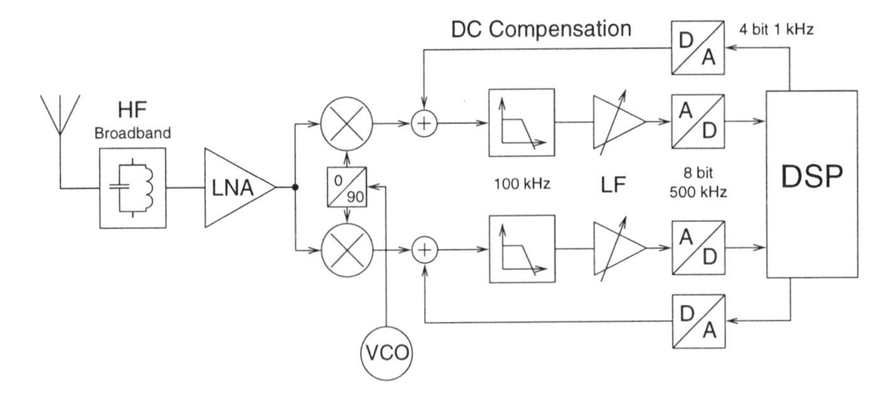

Figure 2.9. Dynamic suppression of parasitic DC-signals in a zero-IF receiver by means of a DSP.

thermore, long time constants can not be integrated with an analog implementation technique because of the large capacitor area that would be required.

The principle of zero-IF receivers has been already known for years [Weav IRE56]. It was however the DC-offset problem that have kept the zero-IF receiver from use in practical applications. It is only with the introduction of DSP's that for some specific applications this problem has found a solution [Seven CICC91, Rab ACD93]. In the DSP a complex non-linear algorithm can be realized that determines the DC-level dynamically. This value can then be fed back to the analog part. This is shown in fig. 2.9. In this way saturation of the lowpass filters is prevented and the distortion is kept acceptable.

At high operating frequencies there will be crosstalk between the two inputs of a mixer. This is caused by capacitive coupling or is, in a fully balanced mixer where the differential structure reduces the crosstalk, due to mismatch in capacitive coupling. The crosstalk between the LO and RF inputs of a mixer results in a DC-voltage caused by multiplication of the LO signal with itself (a_{LO} is the sinusoidal LO signal) :

$$\begin{aligned} & A_{crosstalk} \cdot a_{LO} \cdot a_{LO} \\ A_{crosstalk} \cdot A_{LO} \sin(2\pi f_{LO} t) \cdot & A_{LO} \sin(2\pi f_{LO} t) = \\ A_{crosstalk} \cdot A_{LO}^2 \cdot & \left(\frac{1}{2} + \frac{\sin(4\pi f_{LO} t)}{2} \right) \end{aligned} \tag{2.4}$$

This crosstalk however does not only result in a parasitic DC signal. The multiplication of the RF signal with itself results in broadband parasitic baseband signal (twice the bandwidth of the RF signal). A considerable part of the power of this signal is situated in the lower baseband. This is an unwanted effect because this means that this signal too is superimposed on the wanted baseband signal and once this is done these two signals can not be separated anymore. Fig. 2.10 shows this effect in the frequency domain. It is the result of the convolution of the signal with itself. The actual amplitude and power spectral distribution (PSD) of this parasitic baseband signal depends on the PSD of the RF signal and on whether or not the RF signal is self-correlated. The two cases are discussed.

- *No self-correlation*
  When a signal is not self-correlated it means that there is no correlations between any two frequency components of the signal. In other words : there is no relationship between the amplitude and phase of one frequency component and the amplitude and phase of another frequency component of the signal. All four parameters can vary independent.

- *Total self-correlation*
  Total self-correlation means that there is a relationship between the amplitude and phase of each frequency component. None of them can be varied independently.

An RF signal is, in general, only limited self-correlated. It is self-correlated within small bands. The RF signal is the antenna signal that may already have gone through one filtering step. It has therefore a bandwidth between 10 MHz and 200 MHz. It consists of a collection of signals with a small bandwidth (called channels, with a bandwidth from 200 kHz to 2 MHz) which have total self-correlation, but which are not correlated with each other. The correlation bandwidth of a signal is defined as the maximum space between two correlating frequency components. Frequency components further apart will not correlate. In the case of an RF signal, its correlation bandwidth is equal to the channel bandwidth (or a weighted average of channel bandwidths when channels with different bandwidth are included). A signal with no self-correlation has a correlation bandwidth of 0 Hz.

The power spectral density (PSD) of the square of a signal (caused e.g. by self-mixing) is for a signal with no self-correlation equal to the convolution of the PSD of the sigal with itself. This is not the case for a self-correlated signal. Correlation between its frequency components may either cause an increase or decrease of the PSD of its square. Due to the wanted effect of the self-correlation this will often be an increase. The parasitic signal caused by RF-to-LO crosstalk will therefore be a combination of the two effect. This is shown in fig. 2.11. A large portion of the power will be situated at low frequencies. It can be said that this is the case because the information sources (the channels) within the RF signal are of a low frequency nature. The fact that most of the power of the parasitic signal generated by RF-to-LO crosstalk is situated at low frequencies poses again a problem for the zero-IF receiver whose wanted signal is situated at baseband.

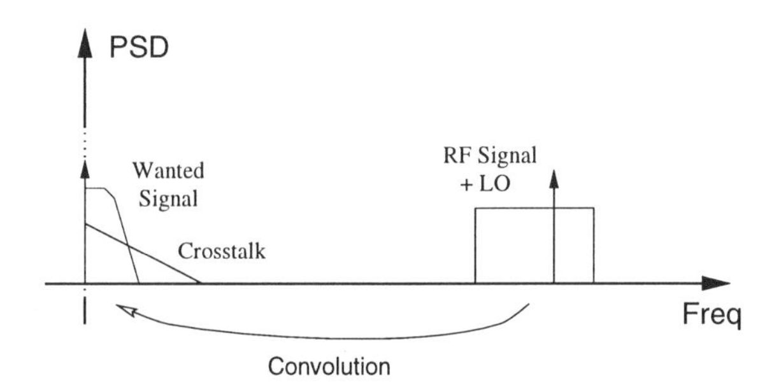

Figure 2.10. The parasitic baseband signal due to crosstalk between the RF and LO input of a high frequency mixer.

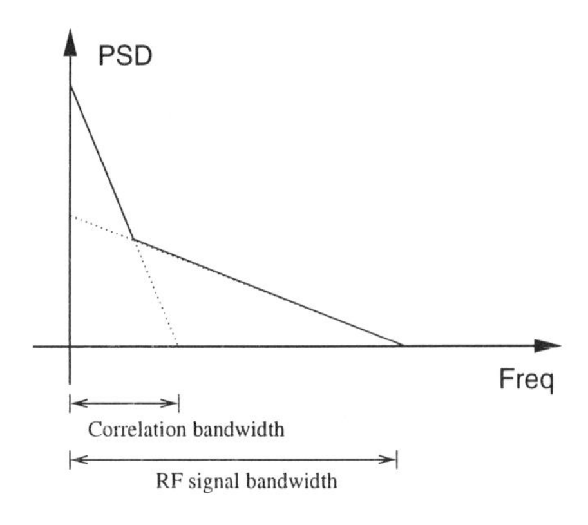

Figure 2.11. The power spectral density of the parasitic baseband signal caused by RF-to-LO crosstalk.

2.4.3 Combined Architectures

The zero-IF receiver topology is, due to its direct downconversion topology, highly sensitive to parasitic baseband signals. It is for this reason that the zero-IF receiver has been used until now in only a very limited number of realizations [Rab ACD93, Min CICC94]. Today, the receiver topology that is most commonly used for newly introduced applications in the 900 MHz to 2 GHz range is the combined IF zero-IF topology [Stetz ISSCC95, Marsh ISSCC95]. The antenna signal is downconverted to a first IF, which is still at a relative high frequency (e.g. from 900 MHz downconverted to 350 MHz), and from there it is further filtered and then directly downconverted with a quadrature downconversion to the baseband. The transceiver shown in fig. 2.3 uses a combined IF zero-IF receiver.

This approach has, compared to the multi-stage IF receivers, the advantage that the number of stages and external filters is reduced. There is only one IF stage and therefore only 2 HF filters and 2 oscillators and mixer stages. It has also the advantage that the output are already quadrature baseband signals so that the wanted signal does not have to be sampled on the IF anymore. Compared to the zero-IF receiver, it has the advantage that, because of the lower frequency from which the direct downconversion is performed, the crosstalk between mixer inputs, caused by capacitive coupling, is reduced. At the same time the wanted signal can be amplified more due to the better filtering at the IF. This gives the wanted signal a higher amplitude and a smaller sensitivity to parasitic signals after downconversion to the baseband.

2.4.4 New Receiver Architectures

The only receiver topologies which are used today are the IF receiver (single or multi-stage), the zero-IF receiver and combinations of both. So, it may seem that, apart from such an obvious but unrealizable solution like direct broadband A/D-conversion, these are the only possible receiver topologies. This is not true. The development of new receiver topologies starts from the perception that it must be possible to realize a receiver which combines the advantages of both known receiver types (the IF and the zero-IF receiver). If one would use two downconversion paths in an IF receiver, all required information for the separation of the wanted signal from the mirror signal would still be available in the two IF signals, as it is in a zero-IF receiver. In this way it must be possible to postpone the mirror suppression from the high frequency part to the intermediate frequency part. In other words : a quadrature downconversion to an IF instead of direct downconversion must be possible. A high quality HF filter for mirror signal suppression would in that case not be necessary anymore and the use of a high IF would also not be required anymore, while at the same time, the receiver would not use baseband operation, eliminating in this way the typical problems of the zero-IF receiver.

Although the concept for new receiver topologies has been proposed in the previous paragraph, it is still not clear how such a receiver can be realized. In [Kaspe PTR83] a receiver which uses a low IF and no high frequency mirror signal suppression was presented. It is however based on the property, typical for FM public radio broadcasting at 100 MHz, that there is never a mirror signal at 150 kHz from the wanted signal and that, at the cost of a lower performance, the bandwidth of an FM signal can be reduced (in this case from 100 kHz to 75 kHz). In this topology there is no mirror signal suppression at low frequency and it brings therefore no real solution to the previously made suggestions. In [Okan TCE92] a receiver topology for FM public radio broadcasting with true mirror signal suppression at a low IF was presented. A discrete realization and the use of a mirror signal suppression technique which is not suited to obtain a high mirror signal suppression lead however again to a limited performance. It is therefore necessary that a new and general theory is developed that analyzes how the information available in both IF signals can be used to separate the wanted signal from the mirror signal after downconversion. The next chapter, chapter 3, introduces the complex signal technique for analog multi-path systems. This technique is a key concept for the analysis and synthesis of receiver topologies. In chapter 3 receiver topologies which use quadrature mirror signal suppression at low frequencies are developed with this method, but also many other new receiver topologies are developed and their advantages and disadvantages are discussed.

2.5 INTEGRATED TRANSMITTERS

2.5.1 The Modulator

The operation of a transmitter can, similar to the receiver, also be split into two parts : the back-end which performs the modulation of the information and the front-end which performs the upconversion of the modulated signal to its carrier frequency. Fig. 2.12 shows this basic configuration. The information, generated in the user-end, is transformed by the modulator into a form which is suited for the wireless communication channel. Again, only digital modulation techniques are considered here as these have become the most widely used for all newly developed wireless applications.

Similar to what was discussed for receivers in section 2.3.2, there are basically two ways in which the digital modulator is interfaced to the upconverting front-end. Either the modulated signals are delivered to the upconverter as quadrature baseband signals or they are converted to the analog domain on an intermediate carrier frequency. The latter technique is called 'direct digital synthesis' [Andre JSSC92].

Fig. 2.13 shows an upconversion front-end which uses direct digital synthesis. The first upconversion, a quadrature upconversion, is performed in the digital domain. The advantage of this technique could, similar to what was discussed for receivers in section 2.4.2, be that there is no baseband operation anymore, making in this way the transmitted signal much less sensitive to parasitic baseband signals. This is not the

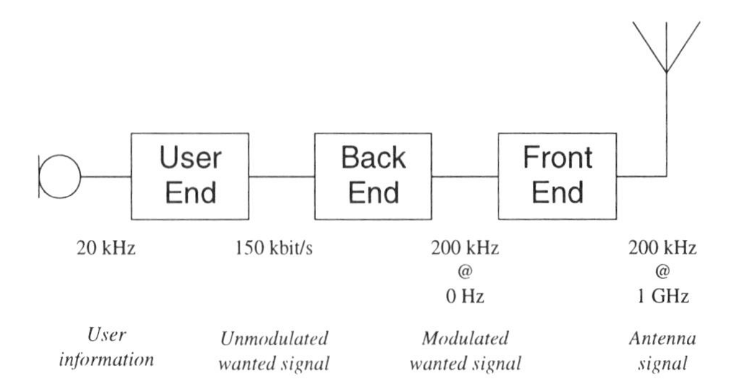

Figure 2.12. The basic configuration of a transmitter for wireless communications; the reverse operation of a receiver.

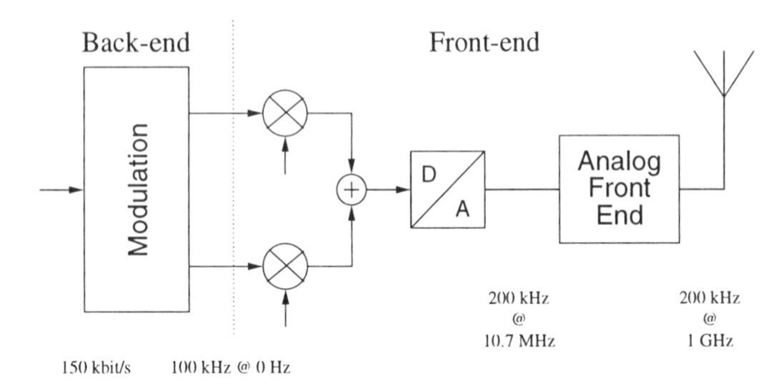

Figure 2.13. A transmitter front-end using direct digital synthesis.

case. The big difference between a receiver and a transmitter is that in a transmitter the only signal present is the wanted signal. The wanted signal is then of course the largest signal present and this imposes only low dynamic range requirements on the transmitter's components. In a receiver the wanted signal can be a very small signal situated next to a very large signal, resulting in the need for a very high dynamic range of its components. The conclusion is that a transmitter is much less sensitive to parasitic signals which are not correlated with the wanted signal. This already indicates why there is no need for the development of new transmitter topologies which avoid baseband operation.

The advantage of direct digital synthesis is that the quadrature upconversion is either performed in the digital domain or that it is performed implicitly in the DSP by using an algorithm which does not require quadrature upconversion. The quality of a quadrature upconversion is highly dependent on the matching between the two paths.

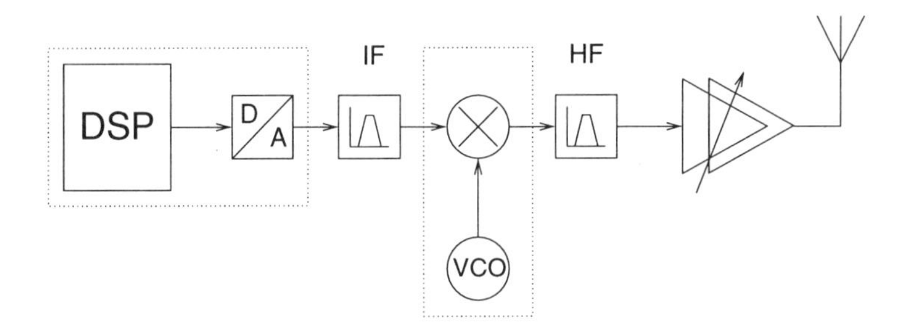

Figure 2.14. The heterodyne topology for transmitters.

It is impossible to achieve perfect matching in the analog domain, while this can be achieved without any problem in the digital domain.

2.5.2 The Upconverter

2.5.2.1 Heterodyne Upconversion. The heterodyne transmitter topology is, even today, still the most often used transmitter topology. Its basic structure is shown in fig. 2.14. The difference between a heterodyne receiver and a heterodyne transmitter is that the transmitter not necessarily requires an IF filter. In the receiver this filter performs the suppression of all adjacent channels, but in a transmitter only the wanted signal is present and whether or not filtering before upconversion is required depends only on the quality with which the modulated signal on the intermediate carrier frequency is generated. A filter on the IF frequency can be used for instance to suppress aliasing products.

The heterodyne upconverter requires thus only a high frequency filter for the suppression of the mirror signal. Such a filter is necessary anyway for any type of transmitter in order to suppress sufficiently out-of-band parasitic signals like harmonic components. The integratability of the heterodyne transmitter therefore would not be so bad. There is however a problem. A single-stage heterodyne transmitter can only be used when the intermediate carrier frequency of the modulated signal generated with the direct digital synthesis technique is high enough. An application working at 900 MHz will for instance require an IF of 200 MHz or more to obtain sufficient suppression of the mirror signal with a standard HF filter. Direct digital synthesis at such a high frequency has been reported [Tan JSSC95], but in general it is very hard to achieve. The only solution which remains is the use of a lower frequency for the DDS and a multi-stage heterodyne upconversion topology. The result is however again a low integratability, due to the need for many off-chip bandpass filters, and a high power consumption, needed to drive these off-chip passive components.

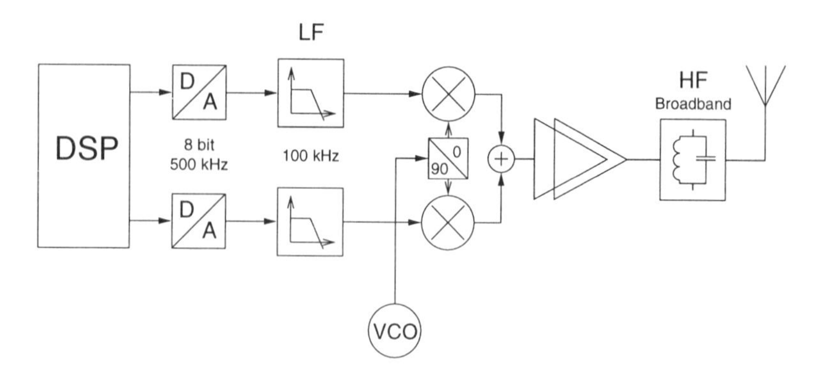

Figure 2.15. The direct upconversion topology for transmitters.

2.5.2.2 Direct Upconversion. The direct upconversion transmitter directly upconverts the modulated baseband signals to their carrier frequency, equal to the local oscillator frequency. Fig. 2.15 shows this topology. The direct upconversion topology can be integrated in a much better way than the classically used heterodyne structure because there is no need for the suppression of any mirror signal generated during upconversion. In the direct upconversion topology the upper and lower sideband of the wanted signal are each others mirror signal and unwanted interference is prevented by doing the upconversion in quadrature.

The disadvantage of the direct upconversion structure is the fact that its local oscillator frequency is equal to the carrier frequency of the wanted signal. This has two drawbacks. First, the crosstalk of the LO signal to the RF output of the upconversion mixers will be transmitted. It is situated at the same frequency as the wanted signal and therefore it can not be suppressed by means of a bandpass filter at the output. The other drawback is the self modulation of the local oscillator that may occur. The wanted signal is upconverted to its carrier frequency and then amplified with a power amplifier to obtain the necessary signal strength for transmission. In some applications the output power of the transmitted signal can be as large as 2 Watt. Such a large signal can easily couple with the sensitive local oscillator when both have the same frequency. The effect will be a of the local oscillator frequency, resulting in a performance loss.

A solution can be found in the use of combined direct upconversion heterodyne topologies [Stetz ISSCC95, Marsh ISSCC95]. In these topologies the baseband signals are directly upconverted to an IF of e.g. 400 MHz and from there they are further upconverted to their wanted carrier frequency. The drawback is a reduced integratability. In chapter 3 new transmitter topologies are developed which combine the advantages of a very good integratability with a good performance, by using a local oscillator frequency significantly different form the wanted signal carrier frequency.

2.6 CONCLUSION

Wireless applications have an important property : they all share the same communication medium, free air. The consequence is that they have to operate in a small passband around a center frequency dedicated to their specific application. This bandpass operation allows for a clear split of the transmitter and receiver functionality in two parts, respectively modulation and upconversion or downconversion and demodulation. The modulation-demodulation (modem) function of a transceiver (transmitter-receiver) has experienced enormous changes with the introduction of digital signal processing. Digital data transmission, digital modulation-demodulation and digital signal flow control has led to high quality wireless applications and an improved frequency re-usability due to cell based operation. The degree of integration of modem and its power consumption has improved significantly and still improves with each introduction of a new technology with lower feature sizes, leading to portable hand-held applications.

The bottleneck in this evolution is the analog front-end which forms the interface between the antenna and the modem. It performs the up- and downconversion. With each newly introduced application the spectrum gets more and more full and apart from the use of cell based systems, the use of ever higher center frequencies is unavoidable. Today, new applications are situated at frequencies between approximately 1 and 2 GHz. These high frequency are a problem for the classically used heterodyne up- and downconverters. In order to achieve a good performance they require more intermediate frequency stages and each stage requires a high quality off-chip bandpass filter and an on-chip driver that consumes a considerable amount of power to drive these off-chip components. Integration is low and the power consumption and cost price stay high. This is the opposite of what is required by the newly emerging wireless applications.

In view of these problems some progress has been made over the last years towards a better integration of transceivers. Zero-IF receivers and direct upconversion transmitters have been developed and implemented. Their operation is based on respectively quadrature upconversion and quadrature downconversion. Both are multi-path topologies which highly depend on matching between two parallel analog signal processing paths. The trend towards integration of several RF signal processing blocks on one chip has made their implementation possible.

Several problems, linked to the fact that they do a direct up- and downconversion, remain however in these topologies and this has limited their use to specific applications which allow a lower performance of the up- and downconverters. The principle of quadrature up- and downconversion and multi-path topologies can however be used for much more than only direct up- and downconversion. What is needed is the developement of a general technique to analyze and synthesize multi-path topologies. In the next chapter such a technique is introduced and used for the development of several new receiver and transmitter topologies which combine the advantage of the classical heterodyne topologies (a high performance), with the advantages of the zero-IF receiver and the direct upconversion transmitter (a very high degree of integration, a low power consumption and a low cost price).

3 TRANSCEIVERS IN THE FREQUENCY DOMAIN

3.1 INTRODUCTION : FILTERING, AMPLIFYING AND FREQUENCY WARPING

A transceiver front-end has to perform three tasks in both the receive and transmit path :

- The center frequency of the modulated wanted signal has to be changed from a low frequency to a very high frequency (in the transmit path) or from the very high frequency to a low frequency (in the receive path).
- All unwanted signals situated outside the wanted signal channel have to be suppressed so that they do not interfere with the correct operation of the wireless communication link and the digital modem.
- The signal levels have to be adjusted in order to obtain the highest possible performance.

The transceiver front-end does not make any change to the shape or form of the modulated signal. This is done in the back-end at low frequencies by means of the modulation and demodulation process. A receiver or transmitter will therefore almost always be realized as a string of operations where each operation is either one of these three frequency domain operations :

- *a filter*, for the suppression of signals outside the wanted signal channel;
- *an amplifier*, to adjust the signal level;
- *a mixer*, to change the center frequency.

Mixers are in this scope actually the combination of both its driving oscillator and its multiplication function. In this way all three functions can be regarded as single-input single-output operators. A receiver and a transmitter are thus both a single chain of operations in the frequency domain without any side branches.

From the previous conclusions it may seem that there are not many transmitter and receiver architectures possible. They must after all be a sequence of one or more times a bandpass filter, followed by an amplifier that re-adjusts the gain, followed by a multiplication with a sine (or a near sine) generated by an oscillator. The design of a transceiver consists then of a proper choice of the bandwidths of the bandpass filters, the gains of the amplifiers and the frequencies of the oscillator. The only architecture choice that can be made is in how many of these stages the up- or downconversion is done. This is the design technique that has been used by most RF designers for the last 50 years and it has led to the very wide spread use of the heterodyne receiver and transmitter topology (both single- and multi-stage).

The heterodyne topology is not the only possible transmitter or receiver architecture. The heterodyne topology is not very well suited for integration and with the increase of the operating frequencies this becomes even worse as more and more intermediate frequencies are required. The zero-IF receiver has been developed [Weav IRE56], especially in the last few years [Seven CICC91, Rofou CICC96], as an alternative receiver topology which does seemingly not consist of the cascade of bandpass filters and mixers typical for heterodyne receivers. Its integratability is much better, but it suffers some major drawbacks and this has limited its widespread use in wireless applications. It is therefore the combined IF zero-IF receiver that gives a good trade-off between integratability and performance which is now the most widely used topology in new transceiver realizations [Stetz ISSCC95, Marsh ISSCC95].

The problem with the zero-IF receiver and the combined IF zero-IF receiver that has been derived from it, is that in these topologies the quadrature approach is only used for direct up- or downconversion. From this the conclusions could be drawn that quadrature signals can only be used in direct conversion topologies. This is not true. Quadrature signal and quadrature operators can be used, just as in a heterodyne topology, as a sequence of several filtering, amplifying and mixing stages. It is even possible to develop new analog signal processing blocks which operate on quadrature signals and which have special properties in the frequency domain that only apply to quadrature signals. In order to gain an insight in these effects, it is necessary that new analysis and design techniques are developed.

In this chapter the complex signal technique is introduced. The complex signal technique is an analysis and synthesis technique which is used in different fields of digital signal processing [Lee 1990, Lee IEEG92, Wack ASSP86]. Its key concept is the use of multi-path signal processing systems which are analyzed as single path systems. Transmitter and receiver topologies which use quadrature signals are an obvious example of such multi-path systems. With the proposed complex signal technique they can be observed as a string of single-input single-output operators with no branches, just like the heterodyne topologies. A logical consequence is that again, sequences of filter, amplifiers and mixers can be built, but now also for multi-path signal processing.

The quality of multi-path signal processing depends heavily on how good the parallel operations in each path are matched with each other. In digital systems this matching can be perfect, in analog systems a perfect matching is impossible. In this chapter, the complex signal technique is therefore extended for the analysis of analog multi-path system and a technique which includes the effects of path mismatch is proposed. The complex signal technique for analog signal processing will be used to analyze the zero-IF topology, the most obvious analog multi-path system. Starting from this discussion several new receiver and transmitter architectures, based on analog multi-path signal processing, will be proposed and analyzed. These newly proposed topologies open the way to a new evolution in transceiver front-end design because their architectures are such that they combine the advantages of both the heterodyne and the zero-IF topology (i.e. a high performance, a very high degree of integration and a low power consumption), rather than to give a good trade-off between them.

3.2 THE COMPLEX SIGNAL APPROACH

3.2.1 Real Signals and their Properties

All electrical signals are real signals. The measured value of a voltage on a given node renders for instance a real number. Measured in time this renders a real signal. The properties of real signals are given in the following equation (with $X(\mathrm{j}\omega)$ the Fourier transform of the signal $x(t)$).

$$x(t) \in I\!R \Leftrightarrow X(\mathrm{j}\omega) = X^{\star}(-\mathrm{j}\omega) \tag{3.1}$$

This means that the negative frequency components of a real signal are always the complex conjugate of the positive frequency components. It is therefore pointless to differentiate for real signals between positive and negative frequencies. There is after all a one-to-one relationship between both. An example :

$$x(t) = \sin(\omega_c t) = \frac{e^{\mathrm{j}\omega_c t} - e^{-\mathrm{j}\omega_c t}}{2\mathrm{j}} \tag{3.2}$$

This real signal has a positive and a negative frequency component. It suffices to know that the amplitude of the positive frequency component is 1/2, while its phase is e.g. 90° . The negative frequency component has, for real signals, always the same amplitude and the opposite phase (in this case -90°).

3.2.2 Complex Signals and their Properties

A complex signal is defined as a pair of two real signals which may be totally independent [Bout RFD89, Lee 1990]. It is a two-dimensional vector of real signals. In most practical cases, this means that a complex signal consists of the voltage waveform on two electrical nodes. Many different representations are possible. The complex representation, with a real and imaginary part, is preferred in this text. The voltage on one node will be assigned to the real part, while the voltage on another node will be the imaginary part. Both signals can be totally independent, although in many applications they will be correlated in one way or another. Practical examples of such complex signal can be found in systems which use multi-path topologies. For instance, the quadrature I and Q signal in zero-IF receivers form a complex signal with I being the real part and Q the imaginary part. In general we have :

$$\begin{aligned} x(t) &\doteq x_1(t) + \mathrm{j}x_2(t) \qquad \text{[by definition]} \\ X(\mathrm{j}\omega) &= X_1(\mathrm{j}\omega) + \mathrm{j}X_2(\mathrm{j}\omega) \end{aligned} \tag{3.3}$$

For example :

$$x(t) = e^{\mathrm{j}\omega_c t} = \cos(\omega_c t) + \mathrm{j}\sin(\omega_c t) \tag{3.4}$$

In this example the complex signal $x(t)$ has only a positive frequency component and no negative frequency component. A complex signal contains the information of two real signals. It contains therefore more information and it is two-dimensional while real signals are one-dimensional. This is reflected in the fact that, different from real signals, for complex signals the positive and negative frequency components can be totally different. Therefore, when using complex signals (in analyzing e.g. multi-path systems) it is important to always make the difference between positive and negative frequencies. Today, in analog circuit design, signals are only considered as real signals and the difference between positive and negative frequencies is never made.

3.2.3 Polyphase Signals and their Properties

It is always possible to take a number of real signals and handle them as one n-dimensional signal vector. Such a signal vector is, in analog design, called a polyphase signal. Not all combinations are however as interesting for analog design. Apart from the complex, 2 dimensional, signals, only the combination of 4 parallel signals is also used [Ging EC73]. Therefore, often the term 'polyphase signals' is used to indicate a 4 dimensional polyphase signal [Ging EC73].

When used in analog parallel signal processing systems, the 4 dimensional polyphase signals are used as the fully differential version of a complex signal. They are therefore analyzed as a complex signal in which both the real and imaginary signal have a differential and a common mode component. Operations defined on these signals will be designed to process the two differential components, while they will have extra circuitry to suppress any common-mode component applied at its input. In this case the polyphase signal has again only two independent components left.

It is possible to analyze polyphase signals as signal vector with 4 truely independent components on which some special operation can be defined. The use of 4 independent parallel signals may not be a wanted effect in multi-path systems, but some signal processing building blocks may not have any common mode suppression circuitry and in this case the polyphase signal analysis technique gives extra information.

A polyphase signal $x(t)$ is defined as :

$$x(t) \doteq x_1(t) + \mathrm{a}x_2(t) + \mathrm{a}^2x_3(t) + \mathrm{a}^3x_4(t) \tag{3.5}$$

Operations on polyphase signals, like the multiplication, are defined with the property $\mathrm{a}^4 = 1$ (similar to $\mathrm{j}^2 = 1$ for complex signals).

3.3 OPERATIONS ON COMPLEX SIGNALS

The use of the complex notation for signals has by itself no real benefits. It is nothing more than a different way of representing available information. Many other ways of representation are possible. The complex representation gives a clear insight in the positive and negative frequency information content of a pair of signals. It becomes interesting when one defines operations on these signals which conserve the properties of the proposed representation. For complex signals this means that the properties of multiplication and convolution should be preserved. This is true for the complex multiplication($(A+\mathrm{j}B)\cdot(C+\mathrm{j}D) = (AC-BD)+\mathrm{j}(AD+BC)$), which is the reason why the complex notation (defined by $\mathrm{j}^2 = -1$) is preferred for the representation of two-signal pairs.

3.3.1 The Complex Amplifier

A complex amplifier multiplies a complex signal with a constant. The result is again a complex signal. The multiplication factor is a complex number. The complex amplification is a linear operation which preserves the frequency position of information

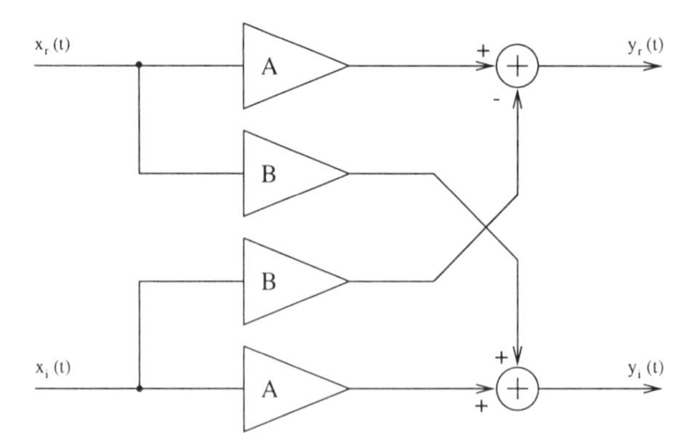

Figure 3.1. Amplification of a complex signal with a complex constant $A + jB$.

available in the complex signal.

$$y(t) = Z \cdot x(t) \text{ with } \begin{cases} y(t) &= y_r(t) + \mathrm{j}y_i(t) \\ x(t) &= x_r(t) + \mathrm{j}x_i(t) \\ Z &= A + \mathrm{j}B \end{cases} \tag{3.6}$$

After substitution this gives :

$$\begin{cases} y_r(t) &= A \cdot x_r(t) - B \cdot x_i(t) \\ y_i(t) &= B \cdot x_r(t) + A \cdot x_i(t) \end{cases} \tag{3.7}$$

The physical realization of a complex amplifier becomes easy starting from equation 3.7. The block diagram is shown in fig. 3.1.

The use of the complex amplifier offers, as such, not many advantages over the normally used real amplifiers. The complex amplifier introduces the same amplitude and phase change to all frequency components of a signal (both positive and negative frequencies). A real amplifier does the same for the amplitude and a constant phase change on all frequency components is, for a stand-alone signal, not important. The amplification with a complex constant is therefore important in feed-back systems, like e.g. complex filters, wherein signals with different phase shifts are compared. This is further discussed in section 3.3.3.

Special cases of the complex amplifier are for instance the real amplification of a complex signal ($Z = A$) and the phase shift over 90° ($Z = \mathrm{j}$) (remark that this gives a +90° phase shift for both positive and negative frequencies). Both cases are shown in fig. 3.2.

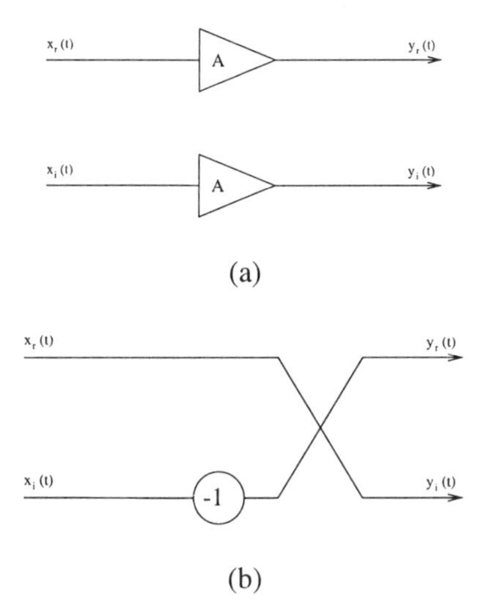

Figure 3.2. Special cases of complex amplification : a) a real amplifier and b) a 90 ^{o} phase shift.

3.3.2 The Complex Mixer

The multiplier is a very important building block for transmitters and receivers because it is used to perform the necessary up- and downconversions. The multiplication of two complex signals can also be physically realized. The implementation complex multiplication,implementation can again be found from the equations.

$$y(t) = z(t) \cdot x(t) \tag{3.8}$$

$$Y(\mathrm{j}\omega) = Z(\mathrm{j}\omega) \otimes X(\mathrm{j}\omega) \tag{3.9}$$

After substitution this gives :

$$\begin{cases} y_r(t) = z_r(t) \cdot x_r(t) - z_i(t) \cdot x_i(t) \\ y_i(t) = z_i(t) \cdot x_r(t) + z_r(t) \cdot x_i(t) \end{cases} \tag{3.10}$$

The realization of this is given in fig. 3.3.

The big advantage of the complex multiplication is that it makes it possible to multiply a signal, complex or real, with a single positive frequency. The consequence is that the problem of the mirror frequency, as illustrated with equation 2.0, does not occur. The mirror signal problem is a side effect introduced by the multiplication with a sine, i.e. the combination of a positive and negative frequency component. With a

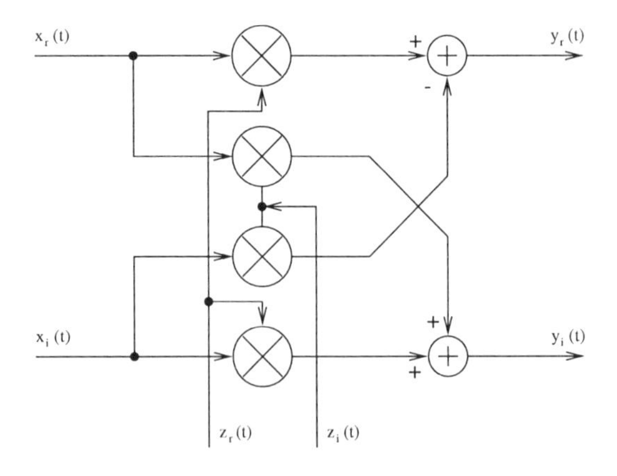

Figure 3.3. Block schematic of a complex mixer.

complex signal $z(t)$

$$\begin{cases} z_r(t) = \cos(\omega_c t) \\ z_i(t) = \sin(\omega_c t) \end{cases} \tag{3.11}$$

$$\Longrightarrow \quad z(t) = e^{j\omega_c t} \tag{3.12}$$

the multiplication becomes :

$$y(t) = x(t) \cdot e^{j\omega_c t} \tag{3.13}$$

In a zero-IF receiver a real signal (the antenna signal) is multiplied with a single positive frequency (the quadrature oscillator signal). This special case of the complex mixer is shown in fig. 3.4.

3.3.3 Complex Filters

A real filter has a real impulse response $h_r(t)$. Its transfer function $H_r(j\omega)$ is a rational polynomial in function of $j\omega$. This rational polynomial can only be real if $H_r(j\omega) = H_r^\star(-j\omega)$. A real filter converts a real signal into a new real signal. On the other hand, the impulse response of a complex filter is a complex signal :

$$h(t) = h_r(t) + jh_i(t) \tag{3.14}$$

$$H(j\omega) = H_r(j\omega) + jH_i(j\omega) \tag{3.15}$$

$H(j\omega)$ is a rational complex polynomial in function of $j\omega$. This means that the coefficients of the polynomial function can be complex. Substituting the equation

$$Y(j\omega) = H(j\omega) \cdot X(j\omega) \tag{3.16}$$

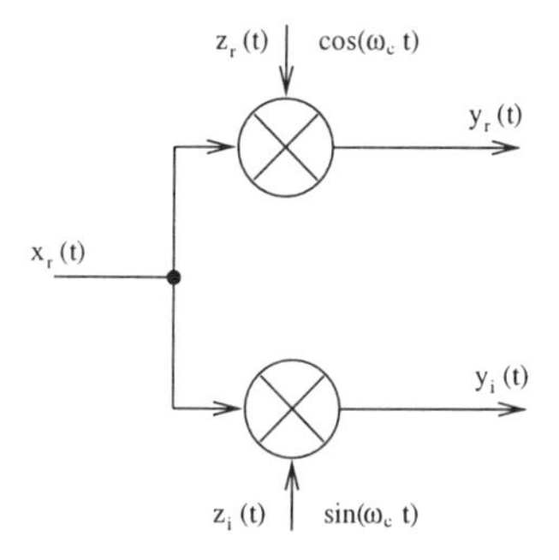

Figure 3.4. The multiplication of a real signal with a complex signal (a positive frequency, $f_c = \omega_c/2\pi$), as used in the zero-IF receiver.

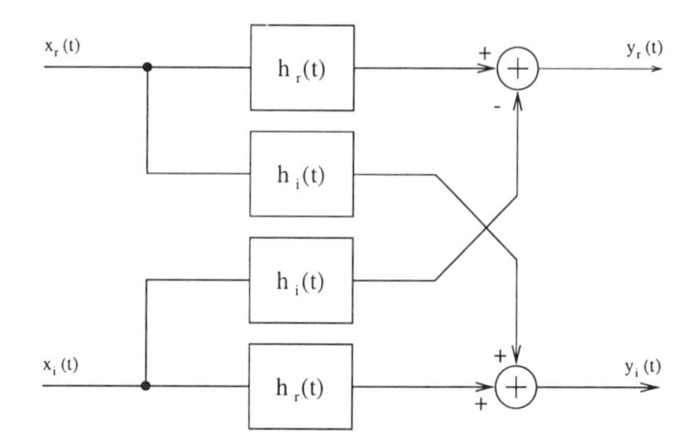

Figure 3.5. General block schematic of a complex filter.

results in :

$$\begin{cases} Y_r(\mathrm{j}\omega) &= H_r(\mathrm{j}\omega) \cdot X_r(\mathrm{j}\omega) - H_i(\mathrm{j}\omega) \cdot X_i(\mathrm{j}\omega) \\ Y_i(\mathrm{j}\omega) &= H_i(\mathrm{j}\omega) \cdot X_r(\mathrm{j}\omega) + H_r(\mathrm{j}\omega) \cdot X_i(\mathrm{j}\omega) \end{cases} \tag{3.17}$$

$$\begin{cases} y_r(t) &= h_r(t) \otimes x_r(t) - h_i(t) \otimes x_i(t) \\ y_i(t) &= h_i(t) \otimes x_r(t) + h_r(t) \otimes x_i(t) \end{cases} \tag{3.18}$$

Fig. 3.5 gives a representation of these equations. A special case is a real filter that filters complex signals. This is shown in fig. 3.6. Another special case is the filtering of a real signal with a complex filter. This results in a complex signal.

3.3.3.1 Linear Frequency Translation. The main application for complex filters is the selective suppression of positive or negative frequency components of a complex

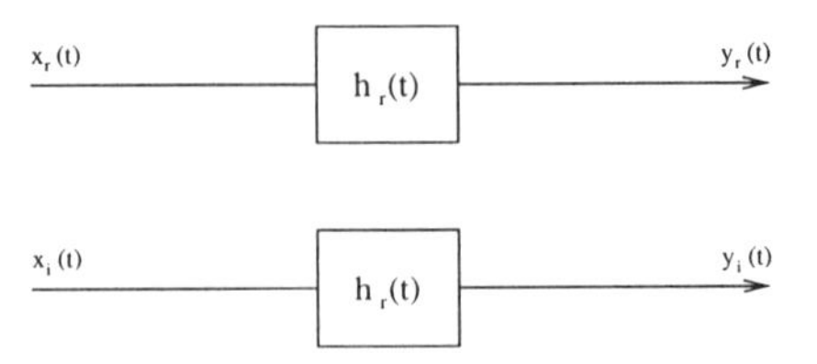

Figure 3.6. A special complex filter : the real filter for complex signals.

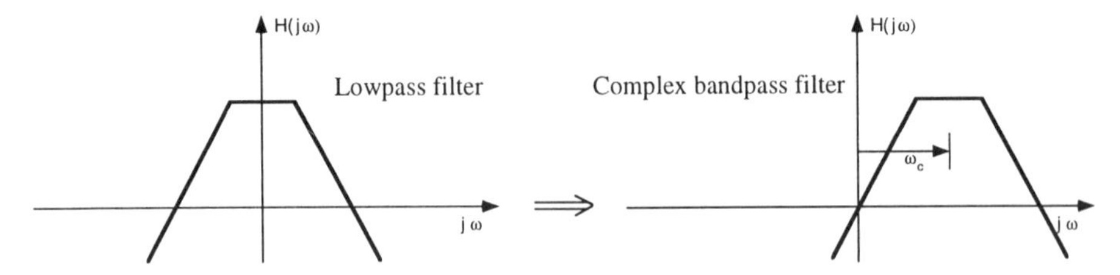

Figure 3.7. A linear lowpass to bandpass transformation, resulting in a complex bandpass filter with a transfer function which passes only positive frequencies.

or real signal [Voorm 1993]. This effect can be realized by using bandpass filters which are obtained by the linear frequency transformation of a lowpass filter. The classical lowpass to bandpass transformations preserve the real properties of the filter

$$j\omega \rightarrow j\omega_c \left(\frac{\omega}{\omega_c} - \frac{\omega_c}{\omega} \right) \tag{3.19}$$

In this case the bandpass filter has the lowpass filter transfer function centered around $\omega \approx \pm\omega_c$. In contrast, the linear transformation is defined as :

$$j\omega \rightarrow j\omega - j\omega_c \tag{3.20}$$

This introduces complex coefficients in the rational polynomial function which implies that it can only be realized with a complex filter. The obtained bandpass filter has only the frequency shifted lowpass filter characteristic for positive frequencies ($\omega \approx +\omega_c$). Fig. 3.7 gives an example of such a transfer function. It shows the suppression of negative frequency components.

The physical implementation of a linear translated lowpass filter could be done with the method of equation 3.17 and fig. 3.5. This is however highly inefficient. For instance, a first-order lowpass filter will be transformed to a second order positive bandpass filter.

$$H_{lp}(j\omega) = \frac{1}{1 + j\omega RC} \rightarrow H_{bp}(j\omega) = \frac{1}{1 + (j\omega - j\omega_c)RC} \tag{3.21}$$

The implementation according to equation 3.17 would require the actual realization of four real second-order filters.

$$\begin{aligned} H_{bp}(j\omega) &= \frac{1}{1 + (j\omega - j\omega_c)RC} \\ &= \frac{1 - j\omega RC}{1 + (\omega - \omega_c)^2 R^2 C^2} + j\frac{\omega_c RC}{1 + (\omega - \omega_c)^2 R^2 C^2} \\ &= H_{bp,r} + jH_{bp,i} \end{aligned} \tag{3.22}$$

This requires thus two second-order filter functions which have to be implemented both twice. The direct synthesis of this filter, discussed below, would only require the realization of two first-order filters. This is even more efficient than the use of a real bandpass filter for the complex signal (equation 3.19). As is shown in fig. 3.6, this would require two second-order filters, while it would give about the same out-of-band suppression, but it would not discriminate between positive and negative frequencies.

3.3.3.2 Direct Synthesis of Complex Filters. The most efficient way to realize the second-order complex bandpass filter of equation 3.21 is with the direct synthesis of the filter transfer function [Voorm USP90, Stey ACD94].

$$\frac{Y(j\omega)}{X(j\omega)} = H_{bp}(j\omega) = \frac{1}{1 - j2Q + j\omega/\omega_o} \qquad \text{with } \omega_o = \frac{1}{RC} \text{ and } Q = \frac{\omega_c}{2\omega_o} \tag{3.23}$$

In this equation ω_o is the bandwidth of the original lowpass filter (before frequency translation), $2\omega_o$ is the bandwidth of the bandpass filter. Q is the quality factor for bandpass filters, defined as the center frequency (ω_c) over the bandwidth ($2\omega_o$). This equation can be rewritten as :

$$(1 - j2Q + j\omega/\omega_o) \cdot Y(j\omega) = X(j\omega) \tag{3.24}$$

$$j\omega/\omega_o \cdot Y(j\omega) = X(j\omega) + (j2Q - 1) \cdot Y(j\omega) \tag{3.25}$$

According to equation 3.25 the filter can be directly synthesized by means of an integrator and a complex amplifier. This is symbolically represented in fig. 3.8. The convention of [Lee 1990] for the represntation of complex signals is used here. The double lines with a single arrow in this figure represent complex signals. The double lines are thus actually representing a pair of two real signals which are treated as if it was one signal. The building blocks can stand for complete complex operators.

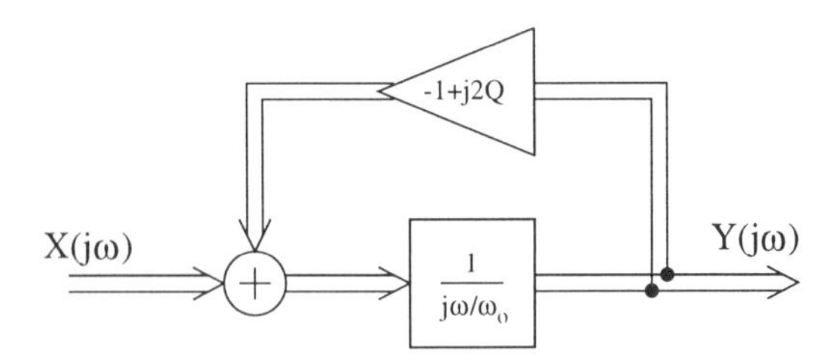

Figure 3.8. Direct synthesis of a second order complex bandpass filter for positive frequencies.

This convention will be used from now on for the compact representation of multi-path circuits. The compact representation can always be translated back in the full implementation with real signals. Fig. 3.9 gives for instance the full block schematic for the actual realization of fig. 3.8. This figure shows that the implementation of the second-order complex bandpass filter with a passband for only positive frequencies indeed requires only two integrators.

The realization of the second-order complex bandpass filter is nothing else than the realization of a linear frequency translated version of a pole. It is also the realization of a complex pole which is not compensated by the realization at the same time of its complex conjugate. Higher order complex bandpass filters can be synthesized by cascading the filter stage of fig. 3.9 several times, once for each frequency translated pole. Fig. 3.10 shows the effect of the linear frequency translation of a 5th order lowpass filter in the s-plane. One real pole and two complex pole pairs are transformed in five uncompensated complex poles. The five uncompensated poles can be realized by using for each pole the filter stage of fig. 3.9. Fig. 3.11 gives the transfer function of a frequency translated 5th order Butterworth filter for complex signals. It has an out-of-band signal suppression which is comparable to the suppression of a 10th order bandpass filter, while it needs only 10 integrators. An extra advantage is that the linear frequency translation gives a perfect lowpass to bandpass filter which preserves the properties of the filter (such as maximally flat amplitude or group delay). A lowpass to bandpass transformation for real filters does not preserve these properties and it gives two passbands, one at positive and one at negative frequencies.

3.3.3.3 Polyphase Filters. The passive sequence asymmetric polyphase filters are another class of analog complex filters [Ging EC73]. Fig. 3.12 shows such a filter with two stages. The main difference between the passive sequence asymmetric polyphase filter and the active complex bandpass filters of the previous section is that they are not bandpass filters, but bandstop filters. They have instead of a passband at only positive (or negative) frequencies, a stopband at only negative (or positive) frequencies.

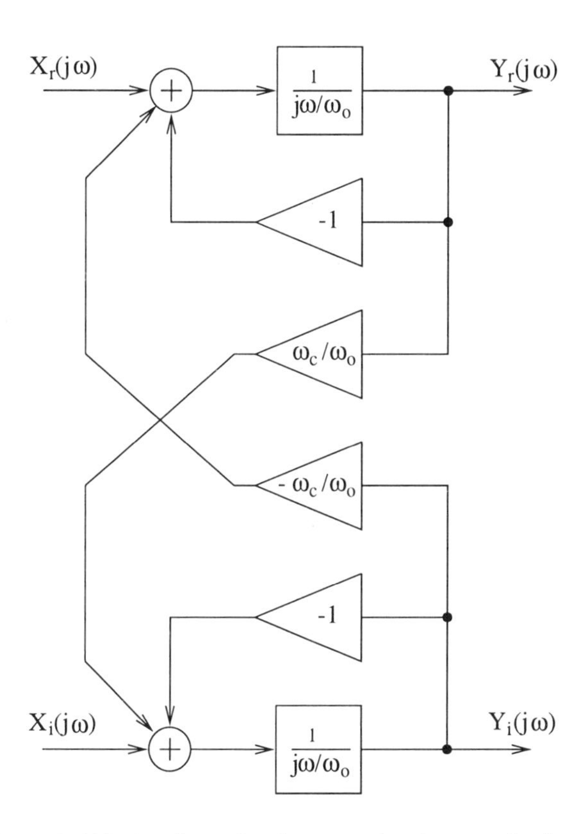

Figure 3.9. Expanded block schematic of a second-order complex bandpass filter.

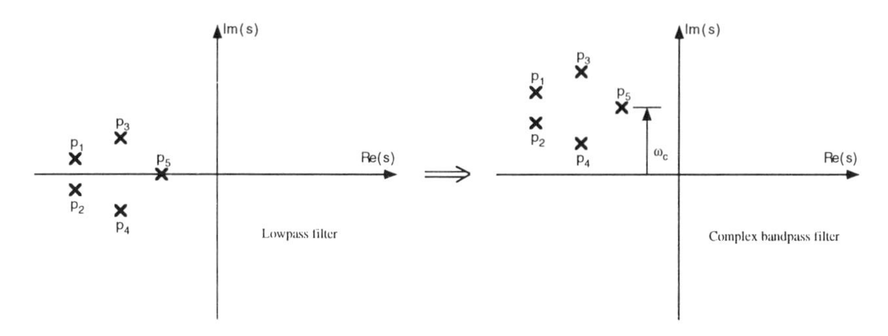

Figure 3.10. The linear frequency translation of a lowpass filter to a bandpass filter in the s-plane.

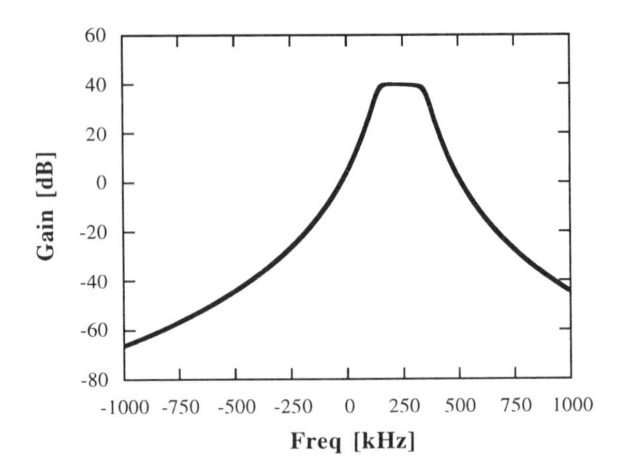

Figure 3.11. The transfer function for positive and negative frequencies of a 5th-order complex bandpass filter.

Fig. 3.13 shows the transfer function of a two-stage polyphase filter for positive and negative frequency components. In comparison with passive sequence asymmetric filters complex bandpass filters are often called 'active integrated sequence asymmetric polyphase filters'. However, in this text the phrasing 'complex bandpass filter' is preferred and passive sequence asymmetric filter are often shortly called 'polyphase filters'. Complex bandpass filters are, because of their active implementation, preferably used in low frequency, low Q implementation, while the passive nature of polyphase filters allows them to be used at very high frequencies.

Passive polyphase filter are actually a symmetric and repetitive version of the classical *RC-CR* allpass filter, often used for the generation of quadrature VCO signals [Stey JSSC92]. Fig. 3.14 shows an *RC-CR* quadrature generator. It can be seen that it gives the correct representation of a single-stage polyphase filter when all inputs but one are connected to the ground node. A polyphase filter stage is a repetitive structure of four times an *RC-CR* allpass filter and each *RC* of one input is shared as a *CR* with another input. The advantage of this repetitive nature is that several polyphase filter stages can be cascaded. By taking the *RC* of each stage slightly different, this allows for the realization of a broad stopband. Fig. 3.13 illustrates this effect.

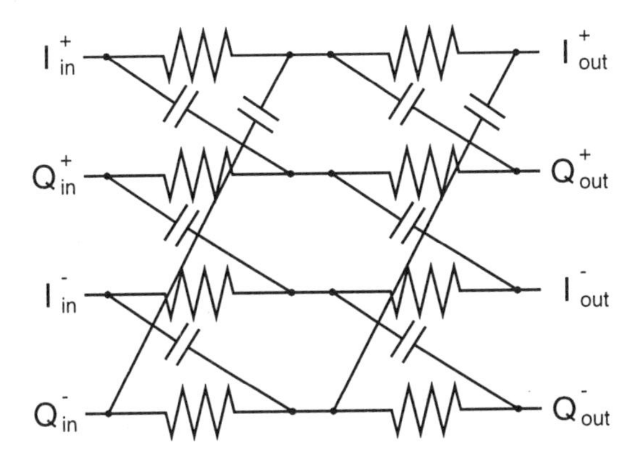

Figure 3.12. A two-stage passive sequence asymmetric polyphase filter.

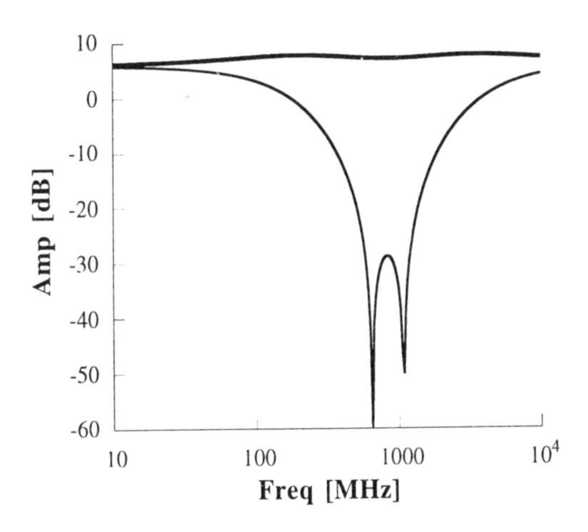

Figure 3.13. The transfer function of a 2-stage polyphase filter for positive frequencies (thick line) and negative frequencies (thin line).

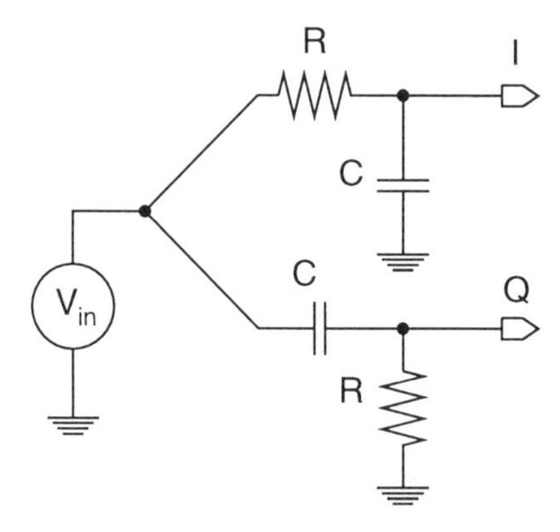

Figure 3.14. An RC-CR filter, classically used for the generation of quadrature oscillator signals.

3.4 COMPLEX OPERATIONS IN THE ANALOG DOMAIN

3.4.1 Mismatch

Complex operators are made with pairs of real operators, amplifiers, mixers and filters. The performance of the system in which complex operators are used degrades when these pairs of real operators are not perfectly matched. In digital systems matching can be perfect and complex operations are widely used [Lee IEEG92, Wack ASSP86]. In analog integrated implementations mismatch between components like transistors, resistors and capacitors is unavoidable.

3.4.2 Frequency Crosstalk

An example of the effects of mismatch on a complex operation is given in fig. 3.15. A perfect amplification of a complex signal with a real constant A does not change the frequency distribution of the signal. The mismatched amplification of fig. 3.15a can, for analysis purposes, be split in four different operations : a perfectly matched amplification with the constant A, an amplification with the mismatched component ΔA, a complex multiplication with the constant $-\mathrm{j}$ and an interchanging of the real and imaginary part of the complex signal. This is shown in fig. 3.15b. The first three operations are complex operations which preserve the frequency distribution. The fourth operation, the interchanging of the real and imaginary part is not an operation which is part of the possible operations defined on complex signals and it does not preserve frequency locations. Interchanging the vector components of a complex signal is equal to reversing the vector sequences. A positive sequence becomes a negative sequence and vice-versa. The interchanging of the vector components of a complex signal is defined in equation 3.26. In equation 3.27 this is written down for a positive and a negative frequency. These equations show that an interchanging of the vector components results in a mirroring of all frequency components around the 0 Hz axis. Here, this effect is called frequency crosstalk (i.e. crosstalk between the information content at positive and negative frequencies).

$$x(t) = x_r(t) + \mathrm{j}x_i(t) \rightarrow y(t) = x_i(t) + \mathrm{j}x_r(t) \qquad (3.26)$$
$$\begin{cases} y_r(t) = x_i(t) \\ y_i(t) = x_r(t) \end{cases}$$

$$\begin{aligned} \mathrm{e}^{\mathrm{j}\omega_c t} &= \cos(\omega_c t) + \mathrm{j}\cdot\sin(\omega_c t) \quad \rightarrow \quad \sin(\omega_c t) + \mathrm{j}\cdot\cos(\omega_c t) = \mathrm{j}\cdot\mathrm{e}^{-\mathrm{j}\omega_c t} \\ \mathrm{e}^{-\mathrm{j}\omega_c t} &= \cos(\omega_c t) - \mathrm{j}\cdot\sin(\omega_c t) \quad \rightarrow \quad \sin(\omega_c t) - \mathrm{j}\cdot\cos(\omega_c t) = -\mathrm{j}\cdot\mathrm{e}^{\mathrm{j}\omega_c t} \end{aligned} \qquad (3.27)$$

With the interchange operation the total transfer function for the mismatched amplifier of the example of fig. 3.15 can be calculated. An applied frequency component will be magnified with a factor A, but it will also be mirrored to its opposite frequency and magnified with a factor ΔA. The superposition principle for linear systems is still valid. The ratio between the unwanted mirrored signal and the wanted original signal will therefore be $\Delta A/A$. An amplitude error of 1 % gives a -40 dB crosstalk between positive and negative frequencies.

The results of a phase error between the two signal paths is also an unwanted mirroring of signals to their opposite frequencies. Equation 3.28 gives the effect of a phase error on a positive frequency component.

$$\begin{aligned} \mathrm{e}^{\mathrm{j}\omega_c t} = \cos(\omega_c t) + \mathrm{j}\sin(\omega_c t) \quad \rightarrow \quad & \cos(\omega_c t + \Delta\phi) + \mathrm{j}\cdot\sin(\omega_c t - \Delta\phi) \\ &= \cos(\Delta\phi)\cdot\left(\mathrm{e}^{\mathrm{j}\omega_c t} - \mathrm{j}\cdot\tan(\Delta\phi)\cdot\mathrm{e}^{-\mathrm{j}\omega_c t}\right) \end{aligned} \qquad (3.28)$$

Apart from the wanted frequency, the total transfer function generates again also a negative and unwanted frequency component. The ratio between the unwanted and the wanted signal is $\tan(\Delta\phi)$, which can be taken equal the $\Delta\phi$ for small values of $\Delta\phi$. This means that for instance a phase error of 1° results in a -35 dB crosstalk between positive and negative frequencies.

3.4.3 Frequency Crosstalk in Analog Implementations

The relationship between component matching and amplitude and phase errors is illustrated with an example. Fig. 3.16 gives the realization of an active-RC single pole

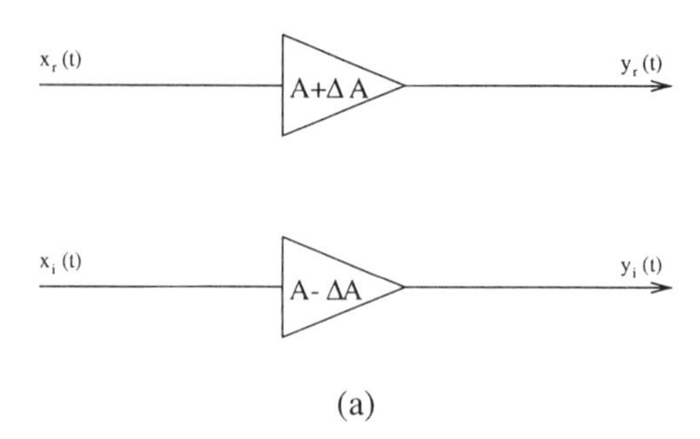

(a)

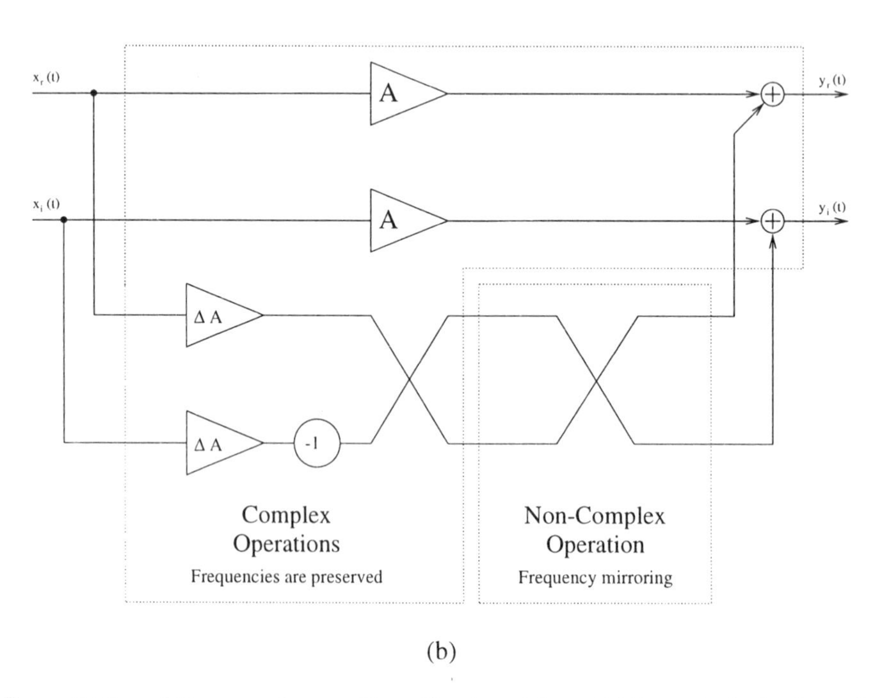

(b)

Figure 3.15. a) The complex amplifier with mismatch and b) its equivalent schematic.

lowpass filter for complex signals. All passive devices differ from their nominal values. There are three sources of unwanted frequency crosstalk in this circuit. One is a constant amplitude error caused by the mismatch on R_1 and R_2. Equation 3.29 gives a calculation and approximation for this error using a worst case analysis.

$$\begin{aligned}
\frac{\Delta A}{A} = \left|\frac{A_r - A_i}{2A}\right| &= \frac{\dfrac{R_2+\Delta R_2}{R_1-\Delta R_1} - \dfrac{R_2-\Delta R_2}{R_1+\Delta R_1}}{2\dfrac{R_2}{R_1}} \\
&= \frac{1}{2}\cdot\left(\frac{1+\Delta R_2/R_2}{1-\Delta R_1/R_1} - \frac{1-\Delta R_2/R_2}{1+\Delta R_1/R_1}\right) \\
&\approx \frac{1}{2}\cdot\left(\left(1+\frac{\Delta R_1}{R_1}\right)\cdot\left(1+\frac{\Delta R_2}{R_2}\right) - \right. \\
&\qquad \left.\left(1-\frac{\Delta R_1}{R_1}\right)\cdot\left(1-\frac{\Delta R_2}{R_2}\right)\right) \\
&\approx \frac{\Delta R_1}{R_1} + \frac{\Delta R_2}{R_2}
\end{aligned} \tag{3.29}$$

The other two sources of unwanted frequency mirroring are the frequency dependent amplitude and phase error caused by mismatch between the cut-off frequencies of the two lowpass filters. Equations 3.29 gives the amplitude error caused by pole position mismatch in function of frequency for a worst case situation.

$$\begin{aligned}
\frac{\Delta A}{A} &= \frac{\sqrt{\frac{1}{1+(\omega\cdot(R_2+\Delta R_2)\cdot(C+\Delta C))^2}} - \sqrt{\frac{1}{1+(\omega\cdot(R_2-\Delta R_2)\cdot(C-\Delta C))^2}}}{\sqrt{\frac{1}{1+(\omega RC)^2}}} \\
&= \frac{\sqrt{\frac{1}{1+(\omega RC)^2\cdot(1+\Delta R_2/R_2)^2+\Delta C/C)^2}} - \sqrt{\frac{1}{1+(\omega RC)^2\cdot(1-\Delta R_2/R_2)^2\cdot(1-\Delta C/C)^2}}}{\sqrt{\frac{1}{1+(\omega RC)^2}}} \\
&\approx \frac{\sqrt{\frac{1}{1+(\omega RC)^2\cdot(1+2\Delta R_2/R_2+2\Delta C/C)}} - \sqrt{\frac{1}{1+(\omega RC)^2\cdot(1-2\Delta R_2/R_2-2\Delta C/C)}}}{\sqrt{\frac{1}{1+(\omega RC)^2}}}
\end{aligned} \tag{3.30}$$

$$\frac{\Delta A}{A} \approx \frac{\Delta R_2}{4R_2} + \frac{\Delta C}{4C} \qquad [\text{for } \omega RC = 1] \tag{3.31}$$

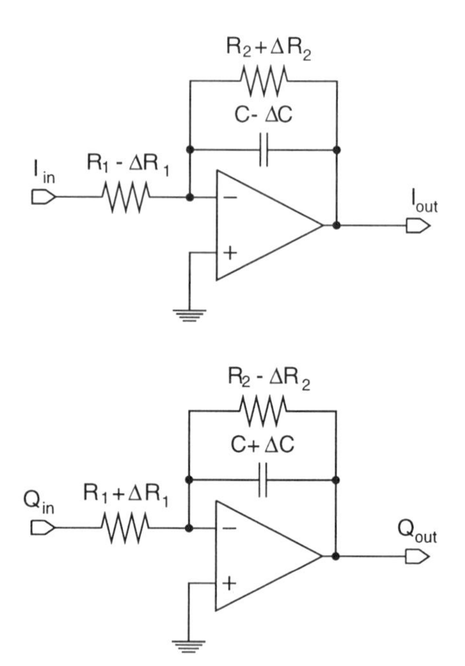

Figure 3.16. Mismatch in a first-order lowpass filter for complex signals.

Equation 3.31 gives the phase error caused by pole mismatch in function of frequency.

$$\begin{aligned}\tan(\Delta\phi) &= \tan\left(\frac{\phi_r - \phi_i}{2}\right) \\ &= \tan\left(\frac{1}{2}\cdot\arctan\left(\omega RC\cdot\left(1+\frac{\Delta R_2}{R_2}\right)\cdot\left(1+\frac{\Delta C}{C}\right)\right)\right. \\ &\quad \left. -\frac{1}{2}\cdot\arctan\left(\omega RC\cdot\left(1-\frac{\Delta R_2}{R_2}\right)\cdot\left(1-\frac{\Delta C}{C}\right)\right)\right) \qquad (3.32)\end{aligned}$$

$$\tan(\Delta\phi) \approx \omega RC\cdot\left(\frac{\Delta R_2}{R_2}+\frac{\Delta C}{C}\right) \qquad [\text{for } \omega RC \ll 1] \qquad (3.33)$$

$$\tan(\Delta\phi) \approx \frac{\Delta R_2}{2R_2}+\frac{\Delta C}{2C} \qquad [\text{for } \omega RC = 1] \qquad (3.34)$$

These three sources of frequency crosstalk are plotted versus frequency in fig. 3.17. The curves of fig. 3.17 are given for a 1 % mismatch on each component. From fig. 3.17 the following conclusions can be drawn :

- Amplitude errors caused by pole position mismatch are small, i.e. capacitor mismatch does not result in amplitude errors.
- Phase errors are caused by resistor and capacitor mismatch. They occur however only at the edge of the passband, meaning that when the filter's passband is larger than the wanted signal band, the phase mismatch is greatly reduced.
- The amplification mismatch is the main cause of signal frequency crosstalk in the passband. It completely depends on resistor matching.

For equations 3.29, 3.29 and 3.31 and fig. 3.17, the worst case analysis technique has been used for the simplicity of the equations and the good insight that can be obtained. An exact statistical analysis of the frequency crosstalk errors of a circuit, its mean values and its variation can be obtained by doing a Monte-Carlo analysis. Fig. 3.18 shows the results of a Monte-Carlo analysis for a linear frequency translated 5th order Butterworth filter implemented with the active-RC technique. A σ_R of 0.1 % on resistor values and a σ_C of 0.2 % on capacitor values is assumed.

3.4.4 Frequency Crosstalk in Complex Operators

In analog implementations perfect matching of components, like resistor, capacitors and transistors, is not possible. It is therefore important to take this into account on the different levels of transceiver design.

- **High-Level Design :** The architecture that is chosen has in several ways an influence on the overall quadrature accuracy that can be achieved due to mismatch in its complex building blocks.

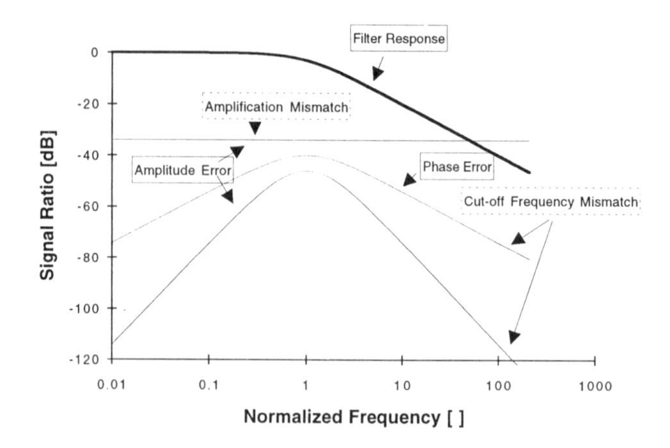

Figure 3.17. Signal frequency crosstalk caused by mismatch in a lowpass filter for complex signals in function of frequency.

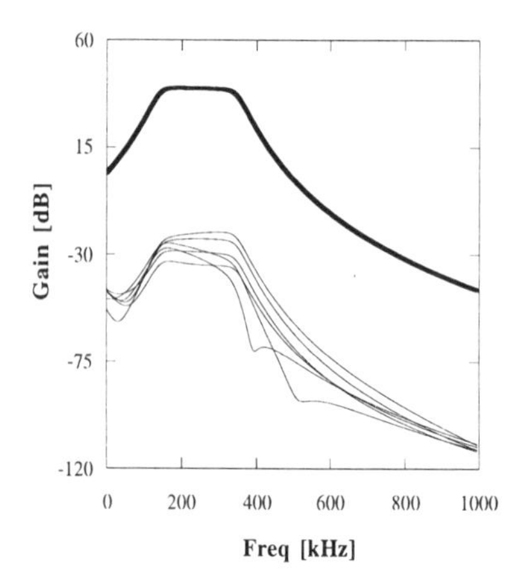

Figure 3.18. Monte-Carlo simulation of the signal crosstalk from negative to positive frequencies for a translated 5th order Butterworth filter (in thin lines; the thick line is the filter's transfer function).

- The architecture itself can be more or less sensitive to quadrature inaccuracies of the building blocks it uses. An example is the double quadrature architecture which will be discussed further on in this text. This architecture is highly insensitive to the mismatch induced frequency crosstalk in its quadrature generators.
- The architecture choice has also an influence on the overall accuracy via the building blocks it requires. Some building blocks are more sensitive to mismatch than others. In a complex filter for instance only the first stage has a significant influence on the crosstalk. In a higher order real bandpass filter, all stages contribute equally, resulting in a higher overall frequency crosstalk. The reason for this is the fact that the unwanted signals at opposite frequencies are already significantly reduced with the filter function of the first stage of a complex filter. A real bandpass filter does not suppress these signals.
- The architecture choice can also influence the matching that can be achieved in the building blocks via the intermediate operating frequencies it requires. Some architectures will require only a good matching at a low intermediate frequency. Building blocks operating at lower frequencies can be designed to have a better matching. This is because larger devices can be used in these circuits.

- **Building Block Design :** For the building blocks it is important to choose an implementation technique that depends on components which have a good matching. For the active-RC implementation resistor and capacitor matching is for instance important, while in case of the *OTA-C* technique this will depend on a good transistor matching.

3.4.5 Corrections of Frequency Crosstalk

In systems where the information content of the complex signal is not random, it is possible to correct phase and amplitude errors. The most obvious example of such signals are the baseband I and Q signal in a GSM receiver. GSM uses GMSK modulation which implies a constant and equal amplitude for both components of the complex signal [Grone TCOM76, Muro TCOM81]. Amplitude errors can therefore be corrected by measuring the amplitude difference between the two parallel signals and adjusting their gain via a high accuracy variable gain amplifier.

The main problem in correcting amplitude and phase errors is to determine their magnitude. For phase and amplitude errors that vary in function of frequency this is almost impossible to do and, in general, corrections are therefore only made for phase and amplitude errors which are constant in function of frequency [Faulk EL91]. In the previous section it was calculated that phase and amplitude errors vary in function of frequency. Their variation is related to the bandwidth of the building block. The bandwidth of low frequency building blocks in transceivers is of the same magnitude as the bandwidth of the wanted signal and their phase and amplitude errors can therefore not be corrected. The bandwidth of high frequency building blocks is much larger than the wanted signal bandwidth and their phase and amplitude appear in the wanted signal band as nearly constant. These errors can be corrected. They will in fact also be the largest errors, because high frequency building blocks have to use smaller devices, resulting in large device mismatch.

A difference between two error correction techniques has to be made : tuning and trimming. Trimming is the technique in which the phase and amplitude errors are measured once, during production. The corrections can be made manually, with externally trimmed resistors or capacitors, on-chip by means of laser trimming resistors or electronically by storing during production a digital code in a ROM or EEPROM and applying this to a DAC [Seven ISSCC94] or a digital correction algorithm. The advantages of trimming is that the signals best suited for the measurement can be applied (typically this will be a sine) and that high quality measurement equipment can be used. Disadvantages are the high cost of the extra production step and the impossibility to correct variations over time of the phase and amplitude (for instance due to temperature effects). Tuning is a technique which continuously, or quasi continuously, measures and corrects the errors during operation. Very low residual errors can be achieved. A lot of digital wireless applications are based on the transmission of data frames. This allows for a fully digital feed forward implementation of the tuning algorithm in the DSP. Phase and amplitude errors are measured per frame and the measured values are then used to correct the data in the frame. A problem is the high complexity of the signal analysis algorithms which fully exploit the properties of the used modulation technique. Currently, the overhead of extra chip area required to implement these algorithms is so high that they are not used for wireless applications. Today, quadrature error correction is mainly limited to trimming techniques, but it may be expected that advanced signal analysis techniques will become more important with the increasing performance of DSP's. In the work that is presented here, the emphasis is on the development of transceiver architectures and building blocks that do not require any form of tuning or trimming, either because they achieve a very high intrinsic quadrature accuracy, or because they do not require a very high quadrature accuracy.

3.5 TRANSCEIVER SYNTHESIS

3.5.1 Receivers

3.5.1.1 The Zero-IF Receiver. The obvious example of a receiver type that uses a multi-path topology is the zero-IF receiver. It can therefore be analyzed with a very good insight via the complex signal analysis technique for analog systems. The topology of a zero-IF receiver is given, in compact notation, in fig. 3.19. The RF signal is

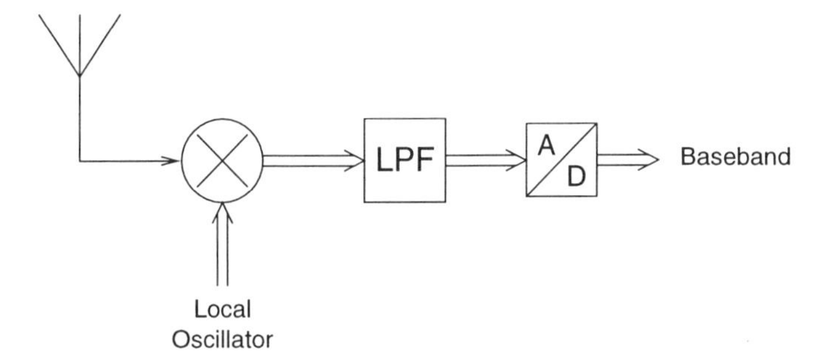

Figure 3.19. The block schematic of the zero-IF receiver in the compact complex representation.

broadband filtered and then downconverted with a single positive frequency equal to the carrier frequency. The broadband filter has to suppress large out-of-band blocking signals, it does not perform any mirror signal suppression. The mixing with a positive frequency makes that only the negative frequency components of the wanted signal are downconverted directly to the baseband and that no mirrored signal is found in the baseband. This process is illustrated with fig. 3.22. In fig. 3.22 the PSD of the input signals and the PSD of the output signal of the downconversion mixer are shown. The arrows between the two PSD representations illustrate the convolution process that takes place in the frequency domain during mixing : one single tone component of the LO signal (white arrow) is mixed with one, either wanted or unwanted, frequency distributed RF signal component (black arrows) and the result is a downconverted frequency distributed output signal. Fig. 3.20 shows for comparison the same downconversion process in an IF receiver.

The zero-IF receiver performs a direct downconversion of the wanted signal to the baseband. The consequence is that the mirror signal is equal to the wanted signal. This does however not mean that there would not be a mirror signal problem in the zero-IF receiver. This would only be true for double sideband modulated signals for which the upper and lower sideband are identical or, in fact, each other's mirror signal. In such a system a downconversion to baseband by multiplication with a single sine would be sufficient (shown in fig 3.21). For all other systems, in which the upper and lower sideband are different, the downconversion must be performed in quadrature. The quadrature downconversion, which is the multiplication with the positive frequency $e^{j\omega_c t}$, makes sure that the upper and lower sideband can still be distinguished at baseband. Matching imperfections in the high frequency quadrature signal and in the quadrature downconversion will cause frequency crosstalk and a suppressed superposition of the upper and lower sideband of the wanted signal. The mirror signal suppression that can be achieved is determined by the amplitude error $\Delta A/A$ and the

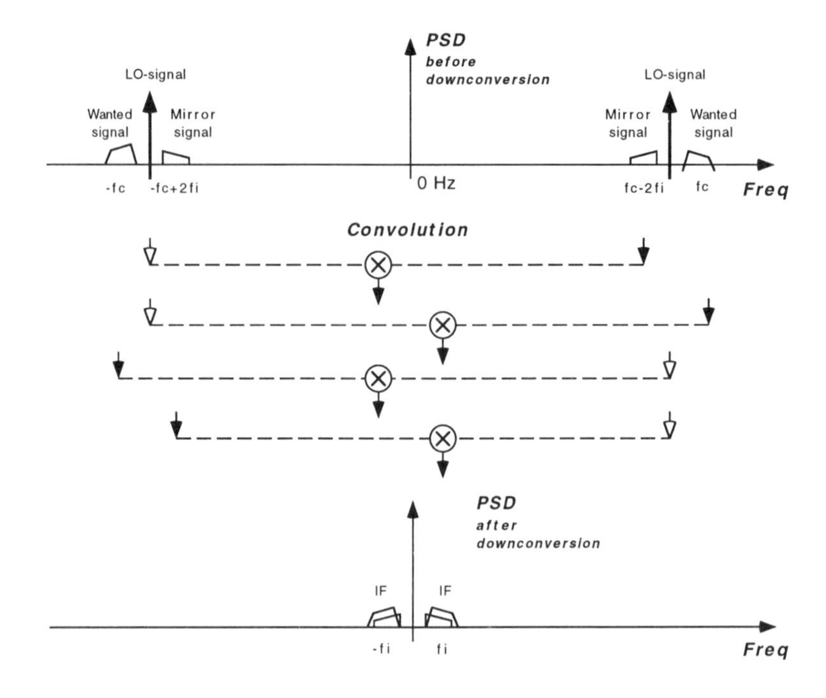

Figure 3.20. The downconversion process in an IF receiver.

phase error $\tan(\Delta\phi)$ of the downconversion mixer and the quadrature signal generator. At high frequencies the quadrature signal error is mainly determined by the phase error. 3° is a typical value for this error, equivalent to a 25 dB mirror signal suppression [Stey JSSC92]. A mirror signal suppression of 25 dB is sufficient for a zero-IF receiver for most applications. In the zero-IF receiver the mirror signal is equal to a mirrored version of the wanted signal and a mirror signal suppression of 25 dB will therefore result in a mirror signal which will after downconversion be 25 dB smaller than the wanted signal. In an IF receiver this would not be true. In the IF receiver the mirror signal can be bigger than the wanted signal, resulting in a higher mirror signal suppression specification for IF receivers.

3.5.1.2 The Low-IF Receiver. The concept of the low-IF receiver starts from the observation that all information necessary to separate the mirror signal from the wanted signal is available in the two low frequency signals after quadrature downconversion [Crols CASII97]. Fig. 3.22 shows that this is true for the zero-IF receiver. Fig. 3.23 shows that this is also true for an IF different from zero. Mixing with a single positive frequency converts only negative frequencies down. In the complex IF signal

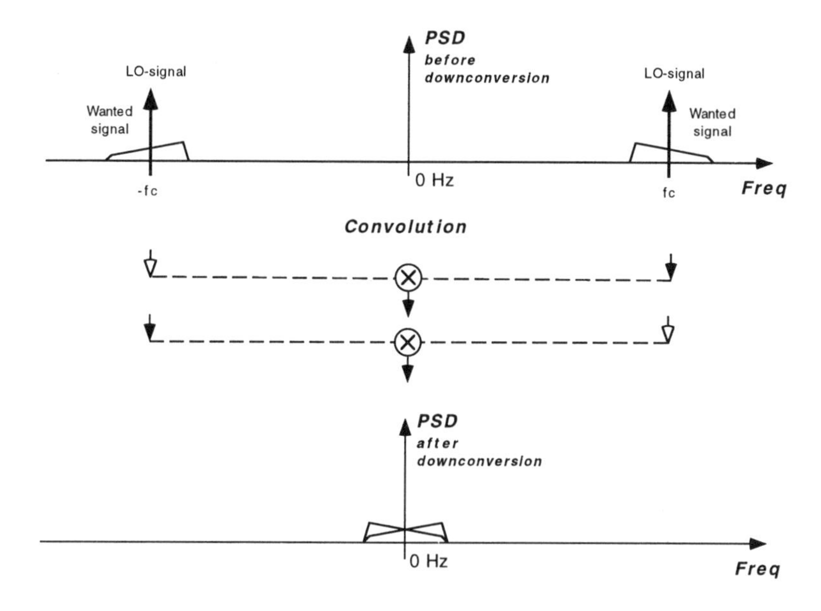

Figure 3.21. The downconversion process in a zero-IF receiver when multiplication with a single sine is used.

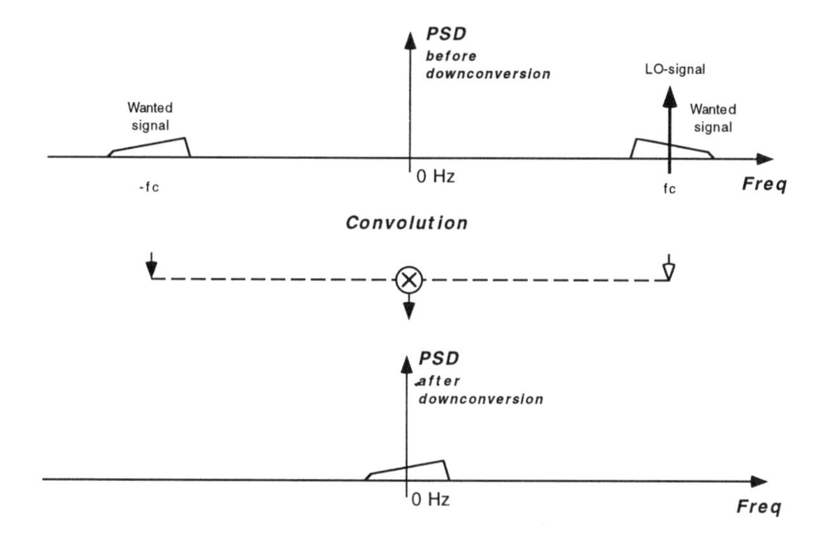

Figure 3.22. The downconversion process in a zero-IF receiver.

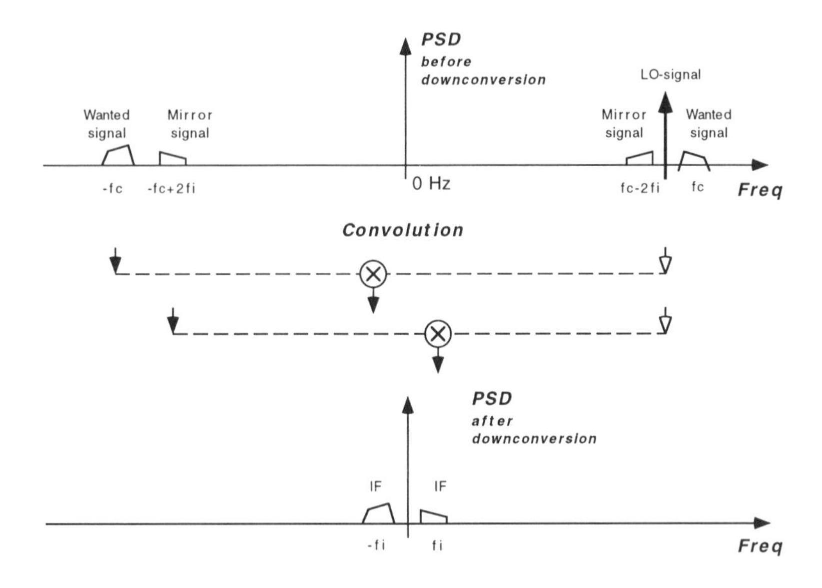

Figure 3.23. The downconversion process in a low-IF receiver.

the wanted signal is now situated at positive frequencies, while the mirror signal is situated at the same, but negative, frequencies.

The downconversion to an IF by multiplication with a single positive frequency has large consequences. They are illustrated with fig. 3.24. It is now not necessary anymore to do any mirror signal suppression at high frequencies before the downconversion. Normally the IF has to be large (10 to 500 MHz) in order to be able to perform the high frequency mirror signal suppression with a bandpass filter with limited Q (< 100). In the low-IF receiver with quadrature downconversion this filter is not required and a low IF frequency can be used (a few hundred kHz, about once to twice the bandwidth of the wanted signal). The result is an integratability that is as good as the integratability of the zero-IF receiver. The zero-IF receiver is actually only a special case (IF = 0) of a receiver with a downconversion by means of multiplication with a positive frequency. The use of the zero-IF makes the topology however highly sensitive to parasitic baseband signals like DC offset voltages, second-order distortion products and self mixing products.

The finite matching between the mixers and, more important, the limited phase and amplitude accuracy that can be achieved in a classical RC-CR quadrature generator, make it impossible to generate a perfect single positive frequency. This may be not so important for a zero-IF receiver, but it is when a quadrature downconversion to an IF frequency. The mirror signal is in that case completely different from the wanted signal

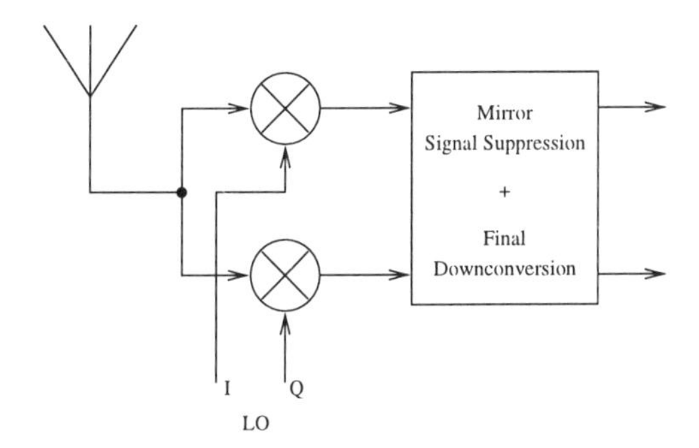

Figure 3.24. The basic schematic of a downconverter for a low-IF receiver topology.

and it can, depending on the application and the exact position of the IF, vary widely in amplitude. Close to the wanted signal it may still have the same amplitude as the wanted signal, while far away it may be more than 50 dB higher than the wanted signal (exact numbers in function of frequency are, for the GSM system, given in chapter 5. In order to realize a low-IF receiver, the use of a quadrature downconverter with a high accuracy must always be combined with a careful choice of the IF frequency.

The accuracy of a quadrature downconverter is too limited and often it is still necessary to do some extra mirror signal suppression at high frequencies before the downconversion. Fig. 3.25 gives two possibilities to do this. A quadrature downconverter combined with a classical HF filter, shown in fig. 3.25a, gives only a limited improvement. Fig. 3.26 gives its operation in the frequency domain. The IF must still be chosen relatively high (e.g. 100 MHz) in order to have a sufficient suppression by the high-Q bandpass filter. Although its requirements may be reduced, the HF filter still can not be integrated. However, this does not have to be a disadvantage because the filter is often already necessary for the suppression of blocking signals. A bigger disadvantage is that a high IF requires a downconversion in several stages (e.g. from 1 GHz to 100 MHz, from 100 MHz to 10 MHz and from 10 MHz to 1 MHz), resulting in a large area and power consumption because of the many mixers and oscillators which are required.

A better alternative which truly combines the advantages of both the IF and the zero-IF receiver, is given in fig. 3.25b [Crols JSSC95b]. A more close observation of the downconversion process by multiplication with a single frequency component shows that it is only the mirror signal situated at negative frequencies that is superimposed on the wanted signal situated at the IF, while the wanted signal is downconverted from positive frequencies. This means that it is not necessary to suppress the mirror signal at both positive and negative frequencies, as is done with the classical high-Q HF filter. This is shown with fig. 3.27. The suppression of only the negative frequency components does not require a high Q, even when the wanted and mirror frequency are

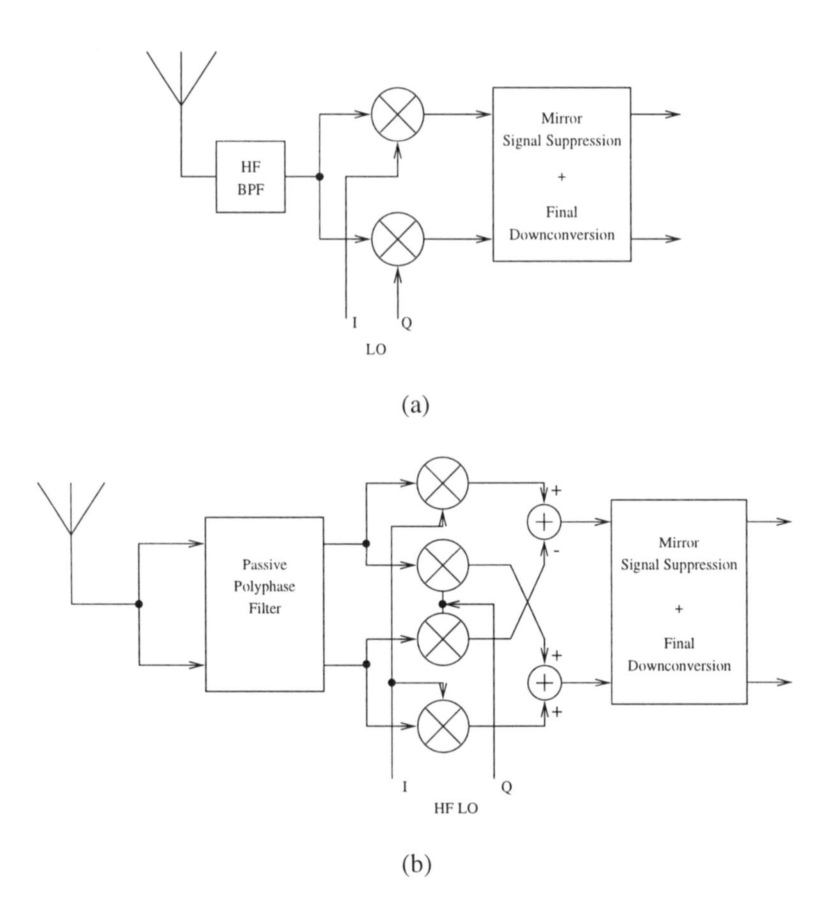

Figure 3.25. Low-IF downconversion with a) an extra external HF filter and b) a sequence asymmetric polyphase filter.

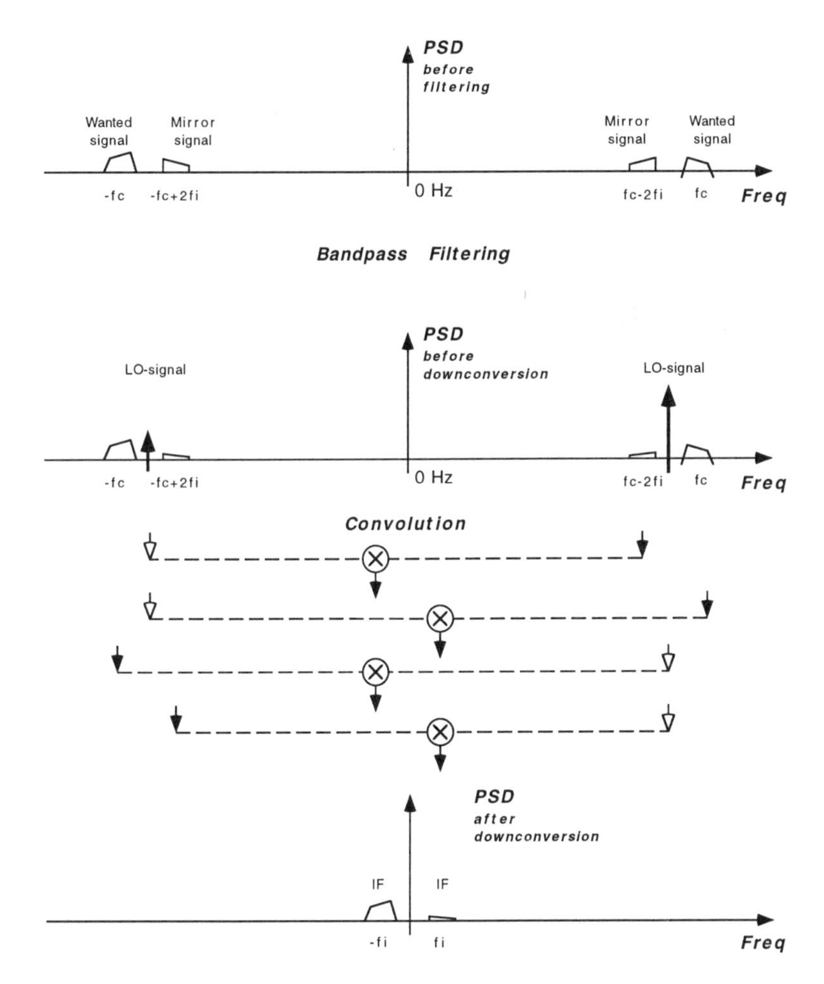

Figure 3.26. Extra suppression of the mirror signal at HF before quadrature downconversion with an external HF filter to compensate for a limited quadrature accuracy of the LO signal.

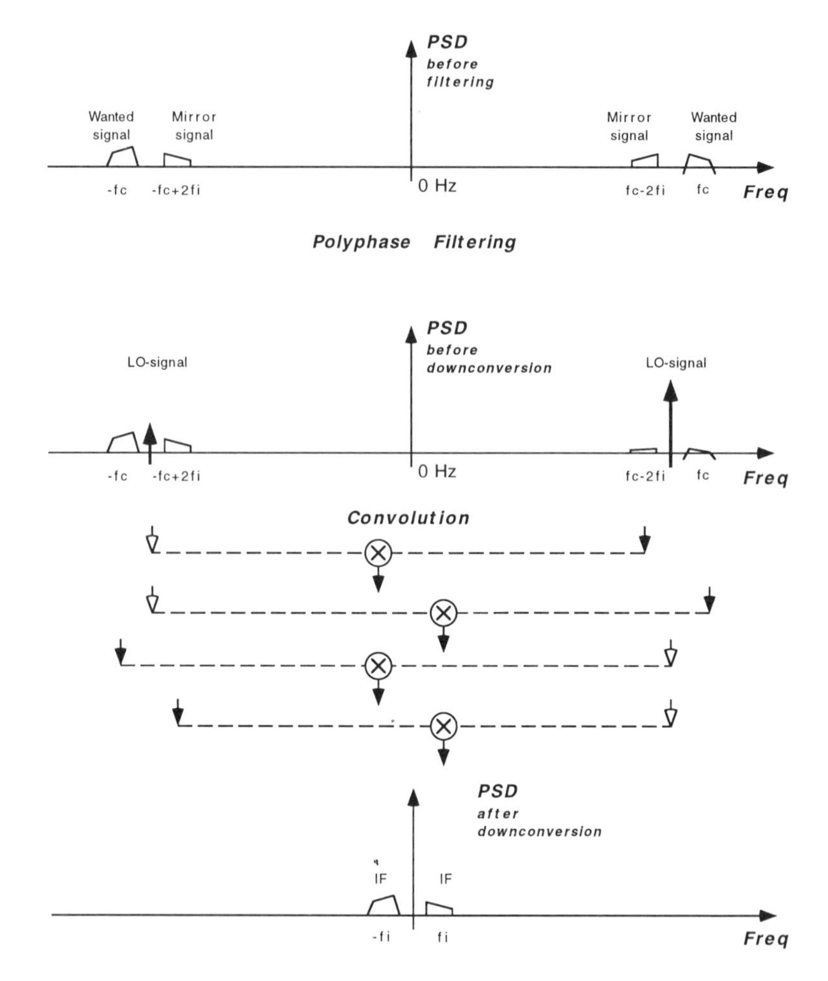

Figure 3.27. Extra suppression of the mirror signal at HF before quadrature down-conversion with a sequence asymmetric polyphase filter to compensate for a limited quadrature accuracy of the LO signal.

situated very close to each other (e.g. a few hundred kHz). The filtering can be done with a sequence asymmetric polyphase filter. Its output signals will of course be an I,Q-signal pair, requiring for the downconversion mixer the full four-mixer structure of fig. 3.3. This newly proposed topology of fig 3.25b is based on the multiplication of two high frequency signals that are both put in quadrature with a quadrature generator. It is therefore called 'double quadrature downconverter'.

3.5.1.3 The Low Frequency Signal Processing. The topology of fig. 3.25b allows for the downconversion in quadrature of the wanted signal to a low IF of a few hundred kHz. In this topology the mirror signal is also downconverted to this same low IF. It is the quadrature structure that ensures that both signals can still be separated after downconversion. The wanted signal is situated at the positive frequency components of the quadrature low IF signal, the mirror signal is situated at negative frequencies. Here is discussed how the mirror signal can be suppressed after the quadrature downconversion from high frequencies and how the wanted signal can be downconverted to the baseband with a final, low frequency, downconversion. In this discussion, it is assumed that the high frequency quadrature downconversion is done with a perfect downconverter.

There are two ways to separate the wanted signal from the mirror signal at low frequencies : the bandpass filtering of only positive frequency components with a complex filter (either active or passive) followed by a quadrature downconversion, or a further downconversion to baseband by multiplication with a single positive frequency component followed by a lowpass filter operation. Fig. 3.28 shows both configurations. In both operations it is the filter operation that largely reduces the dynamic range of the received signal by suppressing the unwanted sideband and mirror signals. A downconversion to baseband can, in the analog domain, only be done with a high accuracy for a low dynamic range signal. High dynamic range signals are too sensitive to parasitic baseband signals. This is the problem of the zero-IF receiver and this is also valid for the topology of fig. 3.28a. The analog version of fig. 3.28a is therefore not a good option. It is however the only possible option when a high IF (e.g. 100 MHz) is used. This is because of the high Q that would be required when a filter is used at a high IF [Gray CICC95]. The circuit of fig. 3.28b renders a signal that can be downconverted to baseband without any difficulty. The dynamic range and the center frequency of the signal after the filter are even so low (e.g. 40 dB and 250 kHz) that it is better sampled at its low IF frequency. This is shown in fig. 3.29b.

The choice between the mirror signal suppression techniques of fig. 3.28a and 3.28b is mainly determined by where the conversion from the analog to the digital domain will be performed. Sampling the low frequency signals before the mirror signal has been suppressed, will require an A/D-converter with a larger dynamic range. As the level of integration and the power efficiency of A/D-converters is continuously improving, allowing for ever more digital signal processing and less low frequency

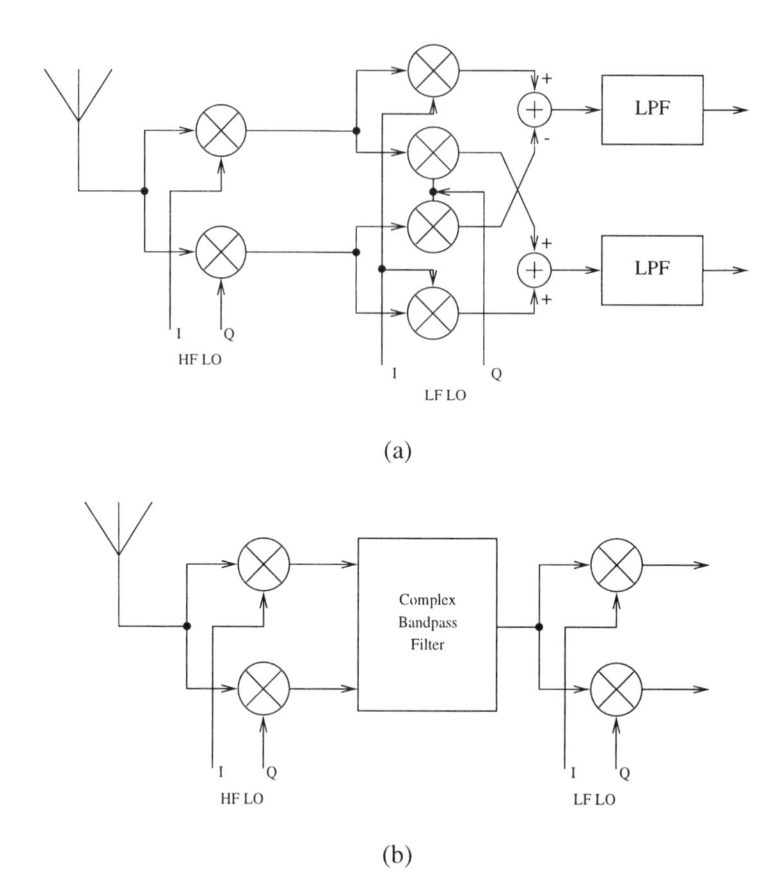

Figure 3.28. Further downconversion from IF to baseband with filtering a) after and b) before the final downconversion.

analog signal processing, the low frequency mirror signal suppression in the digital domain is becoming ever more feasible. Fig. 3.29a shows a topology in which this is the case. The final downconversion and baseband filtering are now done in the DSP with a high accuracy. The analog front-end implements only a very coarse and broadband anti-aliasing filtering. The A/D-converters sample a combination of the wanted signal, the mirror signal and some adjacent channels, resulting in a signal with a dynamic range and bandwidth higher than the wanted signal (e.g. 60 dB and 1 MHz). These are specifications that can be achieved at a reasonable area and power cost with today's embedded A/D-converters. The power cost can even be further reduced by using bandpass $\Sigma\Delta$-converters that only have to achieve the required dynamic range in the band of the wanted signal [Jantz JSSC93, Pelu ACD96]. This makes the topology of fig. 3.29a the preferred topology for the realization of fully integrated high performance receivers. It must of course be used in combination with a highly accurate high frequency quadrature downconverter that does not require the use of a high IF. This can be done with the high frequency downconversion topology that has been presented in fig. 3.25b. Fig. 3.30 shows the full block schematic of the proposed low-IF receiver.

3.5.1.4 The Wideband IF Receiver. The wideband IF receiver is, like the low-IF receiver, also a receiver topology that has recently been introduced as a new receiver topology that combines a high performance with a good integratability [Gray CICC95, Briant ACD96]. Fig. 3.31 shows this receiver topology. For the high frequency downconversion it uses a quadrature downconversion in combination with a high frequency mirror signal suppression filter to give some extra mirror signal suppression (this is the HF topology of fig. 3.25a). The wideband IF receiver uses a fixed frequency downconversion oscillator for the HF downconversion. It downconverts the full application band to a high IF frequency. Channel selection is only done in the second downconversion stage with a variable frequency oscillator. Fixed frequency oscillators have the advantage that a lower phase noise can be obtained due to the lower PLL bandwidth that can be used, allowing on its turn the use of a more noisy voltage controlled oscillator (VCO).

The wideband IF receiver has to use a high IF frequency (e.g. 200 MHz) in order to allow the extra high frequency mirror signal suppression and the downconversion of the full application band. A lower IF frequency would make the relative tuning range of the second oscillator too large. The low frequency downconversion (the second downconversion stage) uses the proposed double quadrature downconversion (see fig. 3.28). Due to the wideband signal at the IF frequency, only a limited amount of lowpass filtering can be used at the IF and the mirror signal will still be present during the low frequency downconversion.

The main advantage of the wideband IF receiver, compared to the zero-IF and low-IF receiver, is that it uses an LO frequency that lies out of the application band, while it

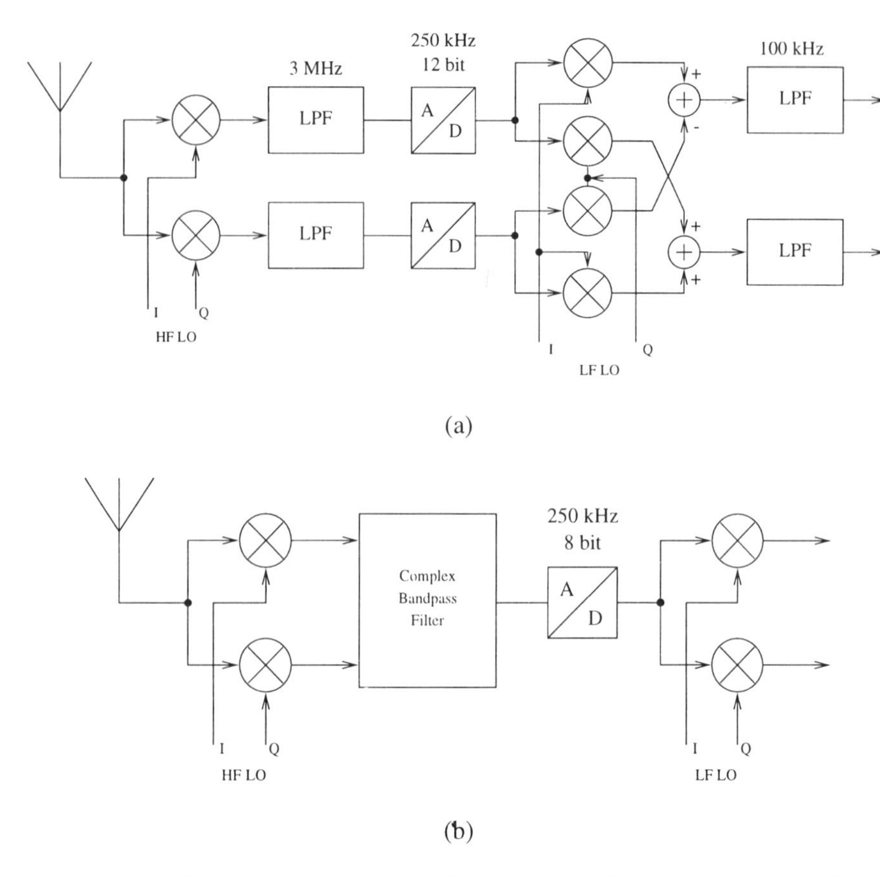

Figure 3.29. Direct sampling of the low-IF signal with filtering a) after and b) before the A/D-conversion.

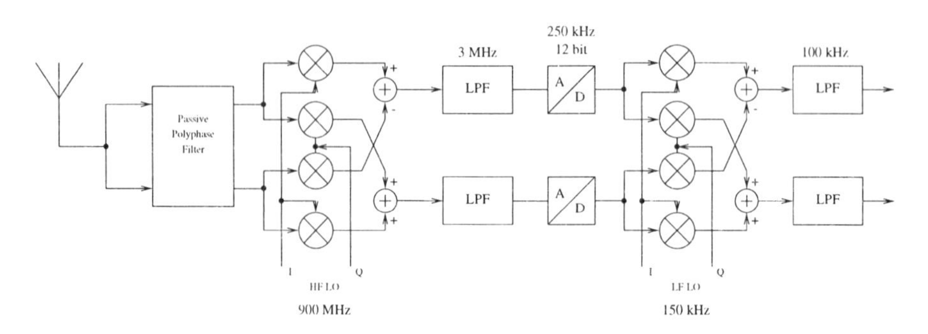

Figure 3.30. The proposed low-IF receiver topology.

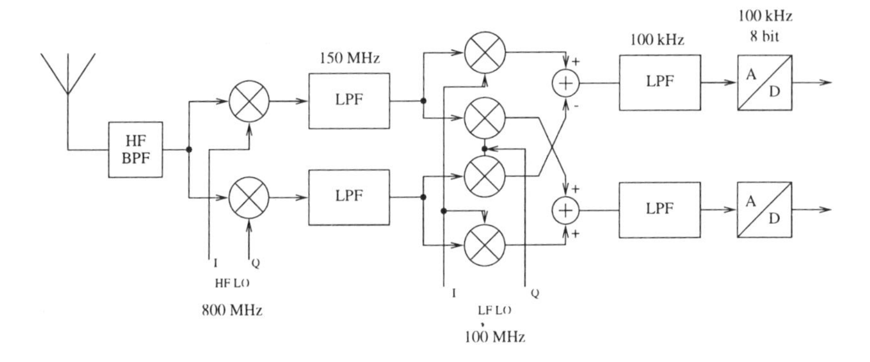

Figure 3.31. The wideband-IF receiver topology.

still offers a very good degree of integratability. In this way the spurious transmission of the LO signal can be reduced. Its main disadvantage is that it performs almost no filtering after the HF downconversion. This makes that the second downconversion stage has to handle high dynamic range signals and all problems common to the zero-IF receiver will be encountered here. The sensitivity to parasitic baseband signals will only be slightly reduced. The wanted signal has not become larger (which is the case in the combined IF zero-IF receiver), but signal crosstalk will be lower at the lower IF frequency. For the second downconversion, the mirror signal problem is larger than in a zero-IF receiver, because here the mirror signal is different from the wanted signal. The HF bandpass filter before the HF downconversion reduces this problem.

3.5.1.5 The wideband IF low-IF receiver. In the wideband IF receiver the mirror signal is already different from the wanted signal. In that case there is no reason why the second downconversion could not be done to a low IF of a few hundred kHz. This is proposed in fig. 3.32. A slightly bigger A/D-converter would be required, but the receiver would not be sensitive to parasitic baseband signals anymore. The mirror signal suppression can be improved by doing a mirror signal filtering before the final downconversion. A bandpass filter is not required for this. A polyphase filter can be integrated and it will give a wideband suppression of the mirror signal which is in this topology situated at the opposite frequencies of the wanted signal. The mirror signal suppression can be so good that a double quadrature downconversion may not be necessary anymore, in this way reducing the required number of mixers and thus the area and power consumption.

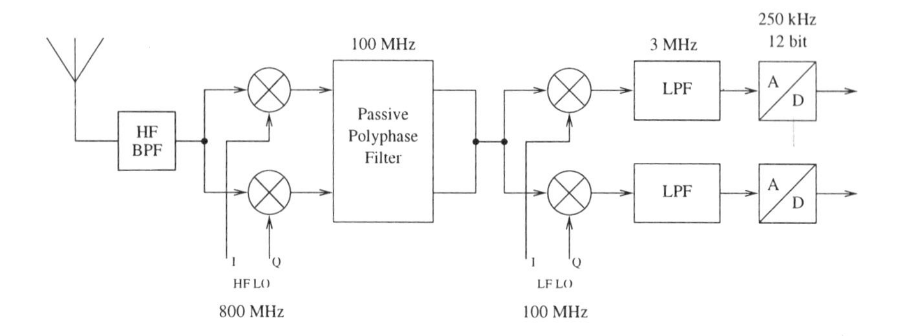

Figure 3.32. The proposed wideband IF low-IF receiver topology.

3.5.2 Transmitters

3.5.2.1 Direct Upconversion. The direct upconversion transmitter topology can be compared with the zero-IF topology for receivers. A quadrature baseband signal is directly upconverted with a quadrature mixer to the wanted carrier frequency. Fig. 3.33 shows the direct upconversion topology in the compact representation for complex signals. In a transmitter the wanted signal is a large signal and sensitivity to baseband signals is therefore not a problem. A limited separation between the upper and lower sideband of the transmitted signal due to a limited quadrature accuracy is a problem that still remains. The main disadvantage of the direct upconversion transmitter topology is however self-modulation. The wanted signal is in the power amplifier boosted to a high power level with the same carrier frequency as the LO frequency. This strong wanted signal can easily crosstalk to the VCO and pull the LO frequency to the frequency of the modulated wanted signal. This is called self-modulation. Self-modulation can be prevented by making a good isolation between the power amplifier output and the sensitive oscillator LC-tank, by using a high loop bandwidth in the PLL or, of course, by using an LO frequency different from the carrier frequency.

3.5.2.2 Double Quadrature Upconversion. The double quadrature downconversion topology, presented in fig. 3.25b, can also be used for upconversion to improve the achievable mirror signal double quadrature upconversion suppression and to reduce the sensitivity to quadrature errors. The double quadrature upconversion topology is shown in fig. 3.34. The quadrature baseband signal is upconverted with a true complex amplifier, resulting in a quadrature high frequency signal. This high frequency quadrature signal is filtered with a polyphase filter. This suppresses the unwanted mirror signal generated by quadrature errors (phase and amplitude errors) on the LO signal. The quadrature accuracy in this structure is therefore reduced and only determined by the matching between the four upconversion mixers.

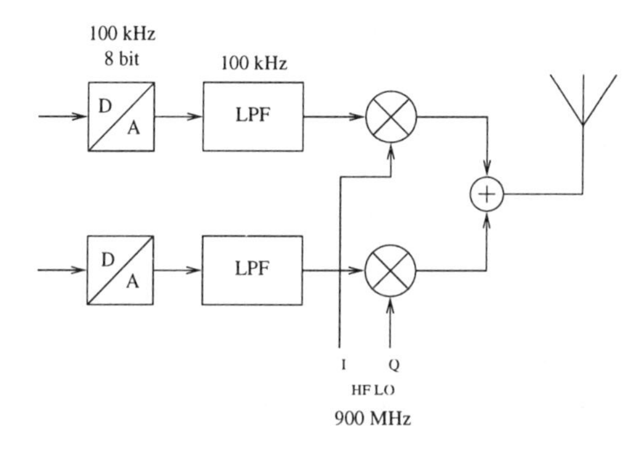

Figure 3.33. The block schematic of the direct upconversion transmitter topology in the compact representation.

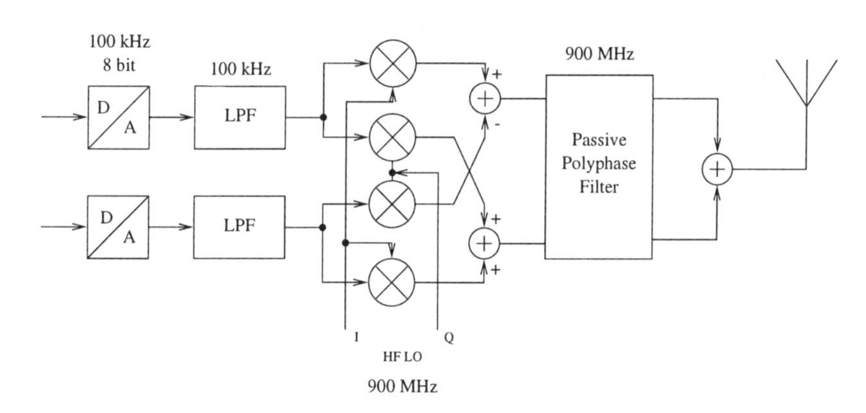

Figure 3.34. The block schematic of the double quadrature upconversion transmitter topology.

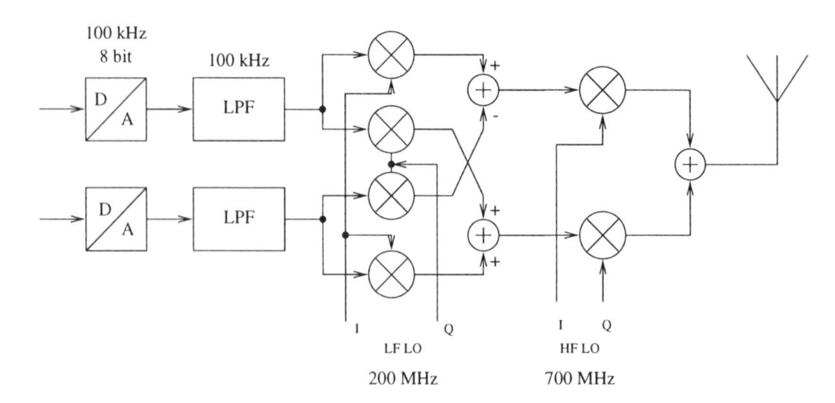

Figure 3.35. The block schematic of a two-stage quadrature upconversion transmitter topology with an IF.

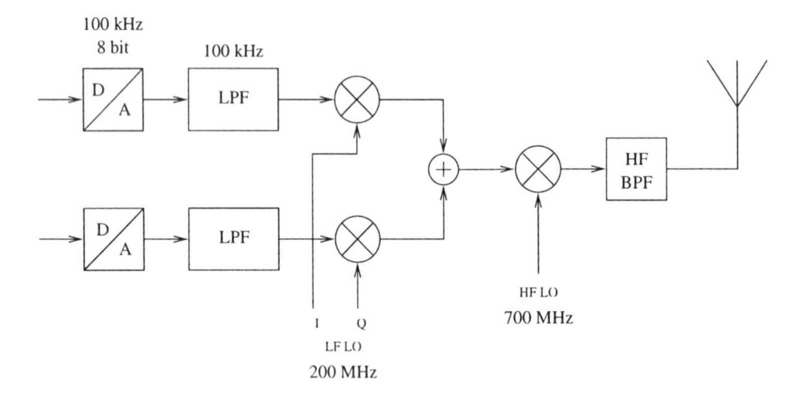

Figure 3.36. The block schematic of a classic IF upconversion topology.

3.5.2.3 Quadrature IF Upconversion. A quadrature IF upconversion topology, as presented in fig. 3.35, can be seen as the equivalent of the wideband IF receiver for upconversion. The quadrature IF upconversion topology has the advantage over the classic IF upconversion topology, shown in fig. 3.36, that no extra HF filter is required after the final upconversion to suppress the unwanted mirror signal. A HF filter will be necessary anyway for any type of transmitter topology after the power amplifier, but the signal level of the mirror would then already be too large to be sufficiently suppressed in order to meet the very high specification on transmission of spurious out-of-band signals.

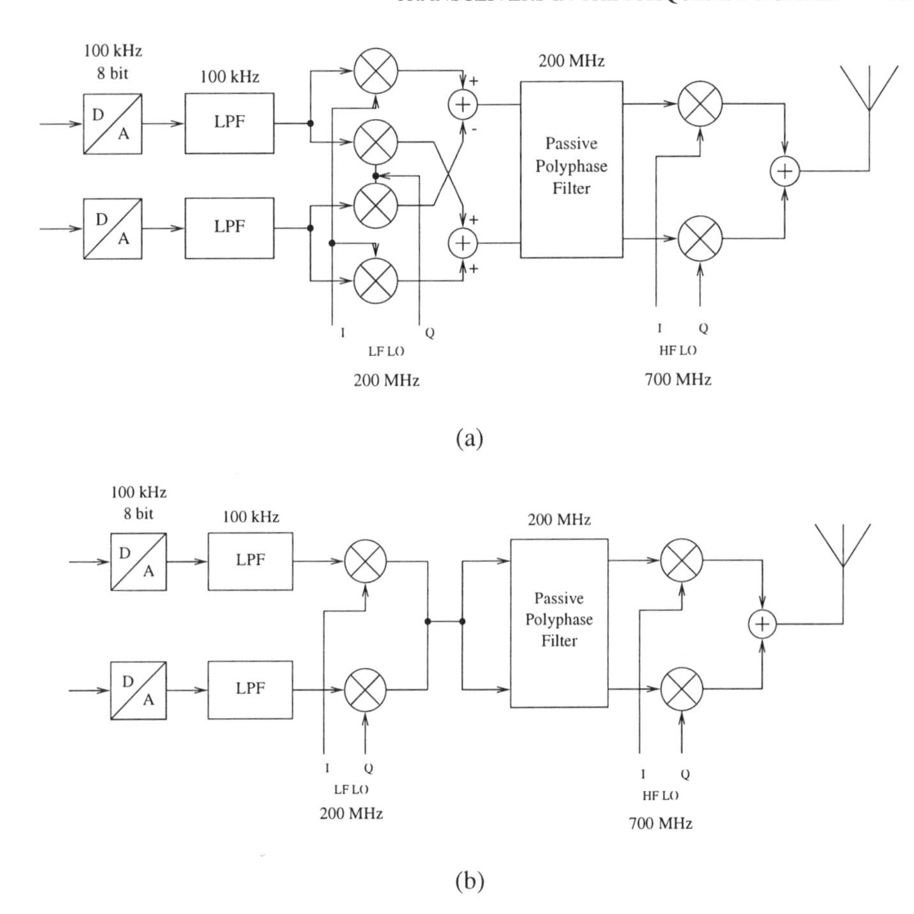

Figure 3.37. Two quadrature IF upconversion topologies which combine a high performance with a high integratability.

3.5.2.4 Special Quadrature IF Upconversion Topologies. Fig. 3.37 shows two variants of the quadrature IF upconversion topology. The polyphase filter in the topology of fig. 3.37a gives an extra suppression of the mirror signal and it compensates in this way the quadrature errors of the first LO. This is the same technique as used in the double quadrature upconverter (fig 3.34). However, the frequency of the first LO is much lower than the operating frequency of the second LO and its quadrature error can therefore be made lower. The double quadrature structure may thus not be required and fig. 3.37b gives an IF upconversion topology which can be fully integrated, which uses a polyphase filter for mirror signal suppression and which uses only a two mixer quadrature upconversion in the first upconversion stage.

3.6 CONCLUSION

The IF (heterodyne) and zero-IF (direct down- and upconversion) topologies are the most often used architectures for the realization of transceiver front-end for wireless applications. Both have their disadvantages (respectively a low integratability and a low performance) and this has led in recent years to the development of the combined IF zero-IF topology which looks for a good compromise between the advantages and disadvantages of the two topologies.

In this chapter it has been demonstrated that many more receiver and transmitter topologies are possible. For this purpose the complex signal technique has been introduced. With this technique, abstraction can be made of the multi-path approach, which is used in, for instance, zero-IF receivers. With the complex signal technique, the analysis and synthesis of multi-path topologies is greatly simplified. Special multi-input multi-output operators, like the complex mixer, the complex filter and the complex amplifier, have been introduced. As a consequence, a whole new range of multi-path based receiver and transmitter topologies could be developed with these techniques. Different from the combined IF zero-IF topologies, these topologies do not compromise between integratability and performance. They combine the property of a very good integratability with a high quality reception and transmission performance.

Essential to the good operation of a multi-path topology is a good matching between the different parallel signal processing paths. In order to be able to use the complex signal technique and multi-path topologies in analog signal processing systems, such as transceiver front-ends, the complex signal technique has been extended. 'Frequency crosstalk' has been introduced as the effect which represents the unwanted operations caused by mismatch in complex and multi-path operators. Sources and effects of phase and amplitude errors can be much better analyzed with this extension of the complex signal technique, even for highly complex transceiver topologies which use many different parallel signal processing paths that are added, subtracted and multiplied with each other.

The general conclusion is that, in this chapter, a new technique has been introduced which broadens the possibilities for receiver and transmitter architecture development significantly. Several new topologies have been presented in this chapter and their advantages and disadvantages have been fully analyzed. These new topologies have such big advantages compared to conventionally designed topologies that it can be expected that the proposed design technique will have an impact on future developments of transceiver architectures for wireless applications.

4 PERFORMANCE OF TRANSCEIVERS

4.1 INTRODUCTION : PERFORMANCE

The front-end of a transceiver converts a modulated wanted signal into an antenna signal, and vice-versa. This is done by means of a sequence of frequency domain operations : upconversion, downconversion, filtering and amplification. Determining which operations are needed and in what sequence is the architecture design. This design will be mainly based on the function that the receiver and transmitter have to perform in the frequency domain. Whether or not a transceiver architecture performs the required frequency shifting and filtering will determine whether or not an architecture is suited. The actual implementation of the transceiver requires the implementation of all the frequency domain operations described by the architecture as physical building blocks, either on-chip or as discrete components. These building blocks will not be perfect. Apart from the wanted frequency domain operations they will also perform some unwanted operations. Unwanted operations are for instance adding noise to the signal and distorting the signal. These unwanted operations will limit the performance of the transmitter and the receiver.

The performance of a wireless link is given by the ratio between the correctly received information and the total amount of transmitted information. A high performance means that most information is transferred correctly. For digital information the performance of the wireless link is often expressed in the form of a bit-error-rate. The performance of the transmitter and the receiver has an influence on the overall performance of the wireless link. Here, the performance of a transmitter or receiver is defined as the ratio between the power of the wanted signal at the output and the total power of all the unwanted signals from which it can not be separated anymore, i.e. the unwanted signals which are located at the same frequencies as the wanted signal. The performance is therefore defined as the output signal-to-unwanted-signal ratio (SUSR). For the transmitter this ratio is taken at the antenna, for the receiver this ratio is taken at its output, before demodulation and after A/D-conversion.

The performance of a transmitter or receiver is determined by the unwanted operations of its building blocks. The magnitude of the unwanted operations of a building block is determined by the building block type (whether it is an amplifier, bandpass filter, mixer, ...), its wanted operation (bandwidth, operating frequency, ...) and two cost parameters : power consumption and chip area. The specifications for the different building blocks, concerning both its wanted and unwanted operations, have to be chosen carefully in order to minimize the cost parameters. This is the high-level design. During the high-level design the specifications given for a certain application are translated in the most optimal specifications for each building block of a given architecture.

In order to do a high-level design it is necessary that the following relationships are known and taken into account :

- All relationships within a building block : specifications of wanted operations, specifications of unwanted operations and cost parameters can not be changed independently.

- The relationships between building block specifications within an architecture and their influence on the total transceiver specifications set by the application.

It is believed that it is very hard to formally describe and quantize all these relationships for all types of building blocks and transceiver architectures. Today, high-level transceiver design is therefore mainly based on experience, re-use of results from previous design and greatly simplifying the design problem. A good feeling for all interdependencies is often only obtained after the lengthy design process has been done all the way down to the transistor level, leaving in this way no room for any optimization. Consequences of this only limited optimization are too high cost parameters (too much power and area consumption) and it has also an influence on the performance that can be achieved. Moreover, the limited knowledge on the relationships between architecture and building blocks makes it very hard to use and select new topologies. A new topology must not only prove to realize the wanted frequency domain operations, it must also achieve the required performance at a very low power and area cost. This will only be known after the high-level design. For a new topology there is no way to do the high-level design and this step is therefore skipped. Low-level design, all the way down to the low-level design transistor sizing, is used to obtain information on the cost parameters for a certain architecture and application. This takes a large amount of time, which makes it therefore impossible to examine, compare and evaluate several possible architectures.

It is important that a good high-level design is done prior to going to the low-level design, because it can significantly reduce the power and area consumption and increase the performance. During high-level design much more can be gained and at a much lower design cost. In order to be able to do a high-level design it is necessary that a technique is developed which allows for a very fast determination of all building block specifications and the cost parameters, without having to go to a lower level of design. In this chapter such a technique is developed. It is based on the development of behavioral models for the transceiver building blocks which do not only describe its wanted and unwanted operations, but which also describe the relationships between wanted operations, unwanted operations and cost parameters. These models are used in a newly developed calculation and optimization technique which translates, for a given transceiver architecture and application, in a very fast way transceiver specifications into the most optimal building block specifications.

4.2 BEHAVIORAL MODELS FOR BUILDING BLOCKS

4.2.1 Modeling of the Building Block Operations

4.2.1.1 Unwanted Operations and Signals. There are many different building block types. They may have different wanted operations (e.g. mixing, filtering, ...), but for the same wanted operation there may also be different implementation techniques for that building block and each one may have different unwanted operations (e.g. a commutating mixer for upconversion has higher components on the harmonics frequencies of the LO signal than a Gilbert type multiplier). The performance of a transceiver will be determined by these unwanted operations. In order to be able to evaluate the performance of a transceiver it is therefore necessary to identify and model all sources of unwanted signals which can be generated by the building blocks. This is done by means of behavioral modeling [Manto AICSP93]. The drawback the behavioral modeling technique is that it requires the development of a different behavioral model for all the different types of building blocks and sometimes even for all implementation techniques. This process is of course very time consuming and research intensive. In this work the possibility of modeling all the different types of building blocks for analog transceiver front-ends with one unique high-level behavioral model is examined.

There are many different types of unwanted operations, often called parasitic effects, which can appear in building blocks. All types of noise, distortion, slewing, ...

To give an exhaustive list of them would be impossible. Nevertheless, it is possible to subdivide all unwanted operations into three categories, according to the type of signal they generate :

- **NOISE :**
  Noise is defined as all signals generated in each building block which are not correlated to any other signal. These are thermal noise and shot noise, but also signals like DC-offset voltages. All parasitic coupling via substrate or power supply to unrelated or quasi unrelated signals (like digital switching signals) will also result in noise type signals.

- **DISTORTION :**
  All signals which only differ in amplitude from a mathematical power of the input signal (different from 1) are called distortion. Main distortion sources are second- and third-order harmonic distortion and intermodulation caused by a non-linear operation, but self-mixing products are also regarded as distortion.

- **ALIASING :**
  Aliasing signals are all frequency translated versions of the input signal that did not undergo the wanted frequency translation. Examples are aliasing components in A/D-converters, but also mirror signals in downconverters and imperfect quadrature downconverters. The effects of phase noise (downconversion of neighbor signals) can also be described as aliasing signals.

The signal processing operation of a building block is thus completely defined by specifying on the one hand its wanted frequency translation (FS, frequency shift) and its linear transfer function (LTF) (i.e. its filter characteristic and amplification) and by specifying on the other hand the (input-referred) noise sources, the distortion levels (mainly second- and third-order) and the unwanted frequency translations (ALIAS). Fig. 4.1 shows how the signal processing in a building block (generating 4 different types of output signals from 1 input signal) can be implemented. Distortion is modeled as distortion on both the input and the output signal. Multi-path signals, like quadrature I and Q signals, can in this formulation be considered as one signal (with complex notation). Some building blocks may, because they incorporate more than one signal processing function, require that they are modeled as a cascade of several signal processing blocks as given in fig. 4.1. For each type of building block a more accurate modeling can also be obtained when each transistor stage in the building block is regarded as a separate signal processing block. Multi-path operators, like e.g. two parallel lowpass filters in the I and Q path, are then considered as one complex operator with a complex signal vector as in- and output. Thus, a receiver or transmitter can in this way be represented for the high-level design as a single path of signal processing. Fig. 4.2 gives, as an example, the architecture for a combined IF zero-IF receiver

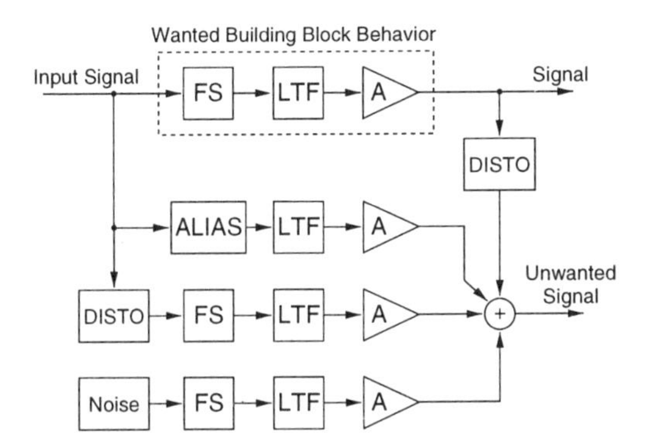

Figure 4.1. The processing of wanted and unwanted signals in a building block.

[Marsh ISSCC95, Stetz ISSCC95] and its representation for high-level design. Upconversion and downconversion mixers must be modeled in combination with their VCO's in order to be able to use the one-input-one-output representation.

4.2.1.2 Unique High-Level Specification of Building Blocks. The wanted operation of any type of transceiver building block can be described with four parameters :

- T_i : The type of building block and its implementation method (e.g. a 5th order lowpass Butterworth filter, realized with the MOSFET-C implementation method).
- BW_i : The bandwidth of the linear transfer function.
- F_i : The wanted operating frequency. For lowpass filters this is 0. For bandpass filters this is their center frequency. For mixers this is the local oscillator frequency.
- A_i : The overall amplification of the building block in its passband. For passive filters A_i is of course equal to 1 or less.

Once these four parameters are chosen, there is only one building block parameter left that can be varied freely.

- DR_i : The noise level, expressed by the input dynamic range which is the ratio between the maximum signal that may be applied at the input and the input referred integrated noise.

With the dynamic range DR_i, the magnitude of all the unwanted signals is specified for an optimally designed building block. An optimally designed building block is a building block in which the relationship between its different unwanted operations is chosen such that its power and/or area consumption (P_i and AR_i) is minimal. It is for instance possible to reduce the distortion of the output stage of an op-amp in

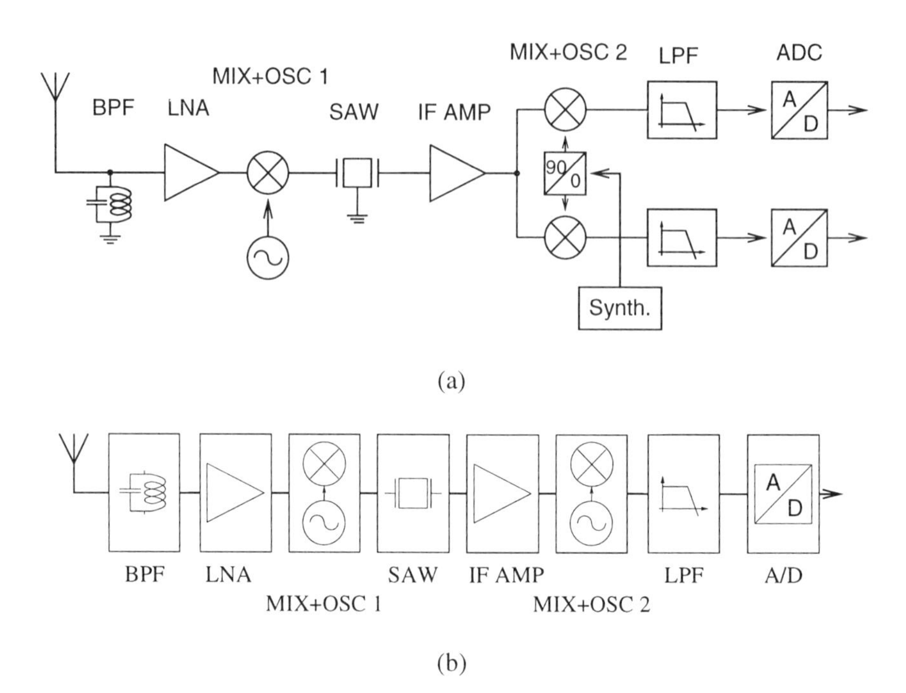

Figure 4.2. a) the architecture of a combined IF/zero-IF receiver and b) its single signal processing path representation for high-level design.

an active integrated filter, but this will significantly increase the power and area consumption while the overall performance of the receiver or transmitter will not change significantly because it then will be the noise of the op-amp which will determine the overall performance. There exists therefore such an optimal design for a given type of building block and a given set of specifications and it may be assumed that only optimally designed building blocks will be used as the possible building block types T_i.

Each building block BB_i is uniquely specified with its five parameters, summed up in the vector $[T_i \ BW_i \ A_i \ DR_i \ F_i]$. They fully describe all wanted and unwanted operations and they also uniquely determine the power and area consumption of the building block. A dependence on technology and the chosen power supply is here of course still present. The types of the building blocks T_i and their sequence in the receiver or transmitter chain has been determined during the architecture choice. The building blocks bandwidths BW_i, operating frequencies F_i, gains A_i and dynamic ranges DR_i have to be determined during the high-level design. During the high-level design these four building block specifications have to be optimized as a function of the SUSR's and performance parameters such as the power consumption P_i and the area consumption AR_i. The power and area consumption of each building block therefore has to be modeled as a function of this parameter set.

4.2.2 Modeling of the Performance / Power Consumption Relationship

In the previous section it was already stated that the power and area consumption, P_i and AR_i, of a building block BB_i are also determined by its set of five parameters and that they can therefore not be used as independent variables during the optimization process of the high level design. In this section this relationship is calculated and modeled for the power consumption of different types of building blocks. Similar hand calculations for the area consumption are not given because they are very hard to derive. Compared to the power consumption, the area consumption of a building block is much more dependent on the type of building block and the used technology and much less on theoretical constraints.

4.2.2.1 Theoretical Minimum Power Consumption. The intrinsic integrated thermal noise power of any signal processing building block with bandwidth BW is $kT \cdot BW$ (with kT the Boltzmann constant, i.e. $4.1 \cdot 10^{-21}$ Joule at room temperature) [Voorm 1993]. The theoretical minimum power consumption is determined by the maximal allowable signal swing S (given in Volts rms) and a dissipating element of the building block, R.

$$P_{min} = \frac{S^2}{R} \tag{4.1}$$

The input dynamic range of the building block is equal to S^2/N^2, with N the equivalent input referred integrated noise in Volt. N^2/R is at least equal to the intrinsic integrated noise power $kT \cdot BW$. This gives for the minimum power consumption :

$$\begin{aligned} P_{min} &= \frac{S^2}{N^2} \cdot kT \cdot BW \\ &= DR \cdot kT \cdot BW \end{aligned} \tag{4.2}$$

The theoretical minimum power consumption of a building block is thus fully determined by its high level design parameters [Groen CAS91, Groen JSSC92, Voorm 1993, Tsivi CICC93]. The efficiency η of a building block is defined as the ratio between its theoretical minimum power consumption and its actual power consumption :

$$\eta = \frac{P_{min}}{P} = DR \cdot \frac{BW \cdot kT}{P} \tag{4.3}$$

4.2.2.2 Modeling of the Efficiency. The performance - power consumption relationship of a real-life building block is accurately described by modeling all its sources of efficiency degradation. The efficiency can be split up as follows :

$$\eta = \eta_i \cdot \eta_s \cdot \eta_p \cdot \eta_n \tag{4.4}$$

By dividing the efficiency in different independent parts, it becomes possible to model all sources of efficiency degradation separately. In this way each physical effect that decreases the efficiency can be analyzed and modeled individually so that it can be used for a wide range of building blocks and building block implementation techniques. The most important physical effects that cause power efficiency degradation are listed in the following paragraphs.

The Intrinsic Efficiency η_i

η_i is defined as the efficiency intrinsically related to the building block type. The efficiency of a certain building block can, even with a theoretically perfect implementation method, never be higher than this value. This factor η_i can be determined for all kinds of filters, mixers, amplifiers and A/D-converters. The most important building blocks are discussed in the following list. Amplifiers must be considered as a special case of lowpass filters.

- **Any type of passive filter, lowpass or bandpass, first and higher order filters :** The intrinsic efficiency η_i of any passive filter is 1. This follows from the definition of the intrinsic integrated thermal noise power from which the definition of η_i has been derived [Voorm 1993].

$$\eta_{i,\text{passive filter}} = 1 \tag{4.5}$$

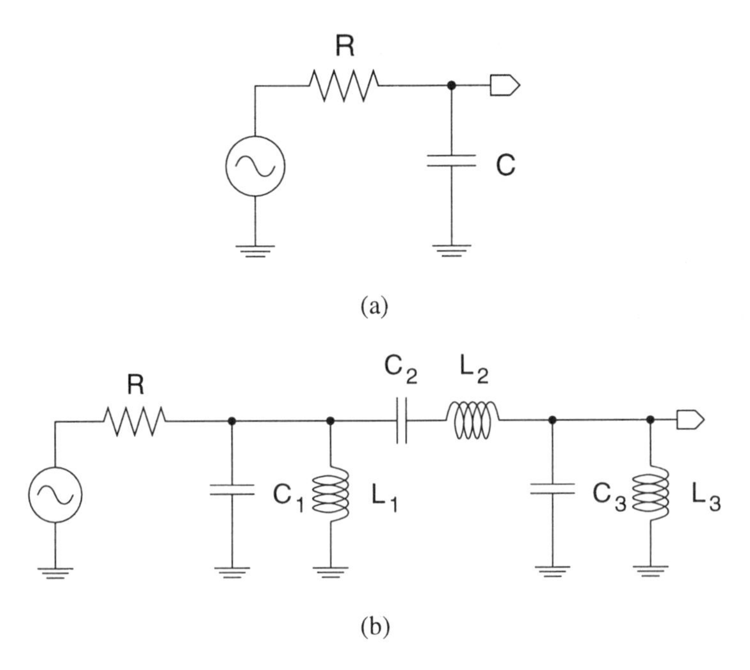

Figure 4.3. Two examples of passive filters, a) a first-order lowpass filter and b) a sixth order bandpass filter.

Fig. 4.3 shows two examples of passive filters. The only dissipating elements are the resistors in the filters and with them the actual filter function is realized. Capacitors and inductors do not generate any noise. The power consumption of these passive filters is the power which the signal source has to drive into the filter.

- **A first-order active lowpass filter with gain** A **:**
  Fig. 4.4 shows the two basic implementation techniques for a first-order active lowpass filter with gain A. In both cases a second dissipating element, A times smaller than R, is necessary. This element has an A times lower noise N^2 than the dissipating element R. The total output noise of an active lowpass filter with gain A is thus $A + 1$ times higher than the output noise of a passive lowpass filter. The input referred noise is than $(A+1)/A^2$ times lower than the input noise of a passive lowpass filter, while the power consumption of the active lowpass filter is A times higher, because it has to drive an A times lower impedance. Combined, this gives

an intrinsic efficiency :

$$\begin{aligned}
\eta_{i,\text{active lowpass filter}} &= \frac{S^2}{N^2_{\text{active lowpass filter}}} \cdot \frac{kT \cdot BW}{P_{\text{active lowpass filter}}} \\
&= \eta_{i,\text{passive filter}} \cdot \frac{N^2_{passive filter}}{N^2_{active LPF}} \cdot \frac{P_{passive filter}}{P_{active LPF}} \\
&= 1 \cdot \frac{A^2}{1+A} \cdot \frac{1}{A} = \frac{A}{1+A} \qquad (4.6)
\end{aligned}$$

The resulting intrinsic efficiency η_i is again about 1. The instrinsic noise and power consumption of the OTA are in the case of an implementation with the OTA-C technique incorporated in equation 4.5. For the opamp in the active-RC technique this is not taken into account because there it can be made negligebly low when a good technology (i.e. with relatively low parasitic capacitances) is used.

Calculations similar to the calculation of equation 4.5 can be made for the other types of building blocks. However, these lengthy calculations are not given here. Instead, only the results of these calculations, their origin and an interpretation of them is given below. The full derivation of these results is left up to the reader.

- **An n-th order active lowpass filter :**

$$\eta_{i,\text{n-th order active lowpass filter}} = \frac{A^2}{(n+A)^2} \qquad (4.7)$$

A higher order active lowpass filter can, for the efficiency analysis, be regarded as a cascade of first-order stages. In reality, complex pole pairs will have to be realized as second-order biquad stages. The quality factor Q of these stages is around 1 (ranging from 0.5 to 2), which gives them about the same efficiency as a cascade of two first-order stages. The use of complex poles instead of real poles will only influence the shape of the frequency domain transfer function around the corner frequency. Fig. 4.5 represents the higher order lowpass filter as a cascade of n first-order lowpass filters. Each stage has the same bandwidth BW, but a different gain. The actual efficiency therefore depends on the division of the total gain A over the different stages. The noise contribution of a given stage is, referred to the input, reduced with the square of the gain of all preceding stages. Therefore, the more the gain is applied in the first stages, the more the efficiency improves. The amount of gain that may be realized in a stage depends on the signal reduction that is realized with the filter function of this stage. A small in-band signal may for instance be accompanied by large out-of-band signals which will be filtered out. This allows

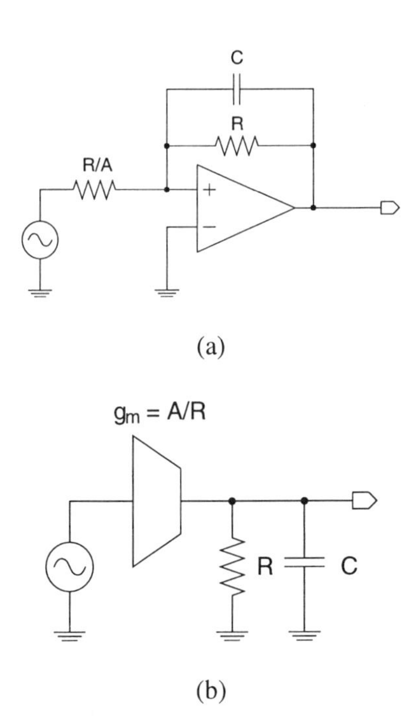

Figure 4.4. The two basic techniques to implement a first-order active lowpass filter are a) the active-RC technique and b) the OTA-C technique.

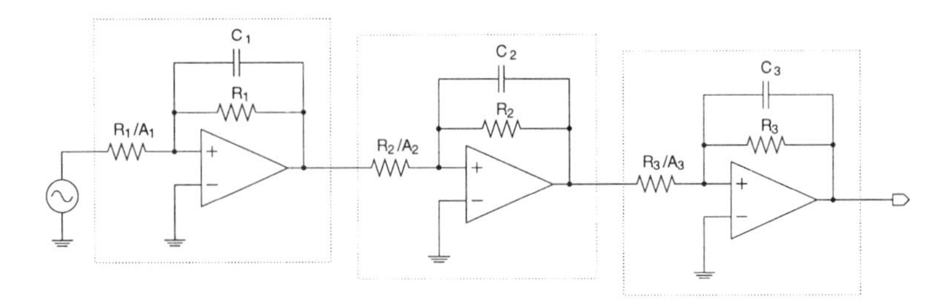

Figure 4.5. A higher order lowpass filter, represented as a cascade of first order lowpass filters with the same bandwidth.

for a large gain. The first stage will give a good suppression of the far away out-of-band signal. This will in most cases result in a large signal reduction, allowing for most of the gain to be realized in the first stage. The following stages will be used to realize a steep roll-off of the transfer function around the corner frequency, resulting in a good suppression of the nearby out-of-band signals. In most cases this will result in less signal reduction. The optimum case is when all the gain is realized in the first stage. This situation is a good approximation for many high gain filters in wireless applications. Once the gain of each stage is known, the efficiency can be calculated. In the optimum case, the noise is $(A+n)/A^2$ times smaller than in a passive version while the power consumption is $A+n$ times higher. The result is again close to 1.

- **A second-order active bandpass filter :**

$$\eta_{i,\mathrm{biquad}} = \frac{A^2}{(1+2Q+A)^2} \tag{4.8}$$

The active implementation of a bandpass filter requires the use of a simulated inductor. Fig. 4.6 shows an example of a high Q biquad with which a second-order active bandpass filter can be realized. A simulated inductor is, different from a passive inductor, not noiseless. Its realization also requires power. Its noise is $2Q$ times higher than the noise generated by the passive element R while its power consumption is also $2Q$ times higher. This gives it a very bad efficiency when Q is higher than the gain A. The gain that can be realized depends again on the applications and its distribution of in-band and out-of-band signals in the frequency domain. For white signals the signal level reduction is equal to the square of the bandwidth reduction. This bandwidth reduction is approximately equal to the quality factor Q and the gain A has therefore about the same magnitude as $\sqrt{Q}$. For high Q filters η_i is thus often much lower than 1. For this reason high frequency

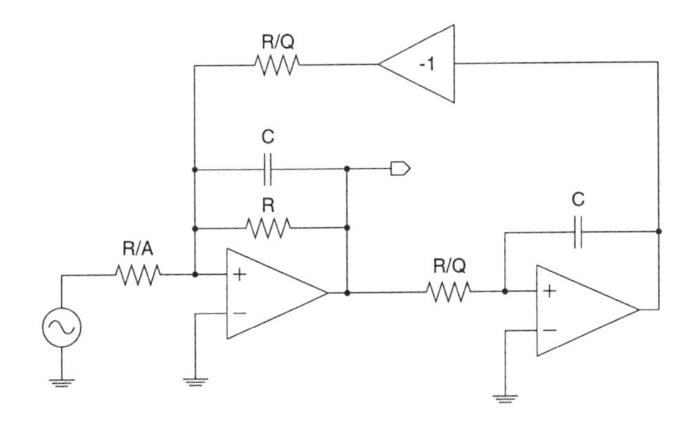

Figure 4.6. A second-order biquad.

bandpass filters (blocking filters, image rejection filters, IF filters, ...) are always passive off-chip components.

- **An n-th order active bandpass filter :**

$$\eta_{i,\text{higher order active bandpass filter}} = \frac{A^2}{(n/2 + nQ + A)^2} \tag{4.9}$$

The efficiency of higher order active bandpass filters can be calculated in the same way as higher order lowpass filters. Most of the gain will be realized in the first biquad stage, but each stage must have the same high Q and each stage will therefore generate a high amount of noise with a high power consumption.

- **A double balanced mixer :**

$$\eta_{i,\text{mixer}} = \left(\frac{G}{A_{amp}}\right)^2 \tag{4.10}$$

A mixer is an intrinsically non-linear building block. Making an accurate noise analysis of a mixer is therefore an elaborate task. Noise may be modulated, it may be mixed up and down and noise sources may vary periodically during time. A dynamic noise analysis is therefore needed. For double balanced mixers (almost all integrated mixers are double balanced) it is nonetheless possible to use a static noise analysis for only white noise which is accurate enough for the behavioral modeling. The reason for this is that the overall impedance of the modulated transconductance or resistive elements stays constant over time. The noise which they produce is therefore also constant in time and white noise that is up- or downconverted still

remains white noise.
A double balanced mixer can be seen as an amplifier from which the gain can be varied between $+A_{amp}$ and $-A_{amp}$. The output noise of the double balanced mixer during operation is equal to the output noise of the mixer when used as amplifier. The actual gain of the amplifier has no influence on this. Because of the double balanced structure the mixer has the same white output noise spectrum under all possible input conditions.
The conversion gain G of a mixer is defined as the ratio between the amplitude of the wanted signal at the output and its amplitude at the input. Remark that here, different from a filter or amplifier, its frequency position may be changed. The conversion gain G depends thus on the amplitude and shape of the oscillator signal. A lower oscillator signal amplitude will result in a lower conversion gain. Mixing with a block signal instead of a sine with the same amplitude will result in a slightly better conversion gain. The conversion gain G is always lower than the maximum possible gain when used as an amplifier A_{amp}. When modulating with a sine with the highest possible signal swing, the ratio between A_{amp} and G is 2 [Crols JSSC95a]. The efficiency of a double balance mixer depends on the ratio G/A_{amp}. The input referred noise of the mixer is equal to the output referred noise divided by the conversion gain G. When used as amplifier this is A_{amp}. The ratio between the two gives the efficiency reduction.

- **A quadrature downconverter :**

$$\eta_{i,\text{quadrature mixer}} = \frac{1}{4} \cdot \left(\frac{G}{A_{amp}} \right)^2 \tag{4.11}$$

A quadrature downconverter with double balanced mixers is a two-input four-output device (the oscillator is modeled as part of the mixer). By using the differential and complex signal analysis the quadrature downconverter can still be regarded as a single-input single-output device. The efficiency of this device is calculated. The I and Q output signal after quadrature downconversion can, with complex signal notation, be described as one signal on which noise that comes from the two mixers is present :

$$a_{out}(t) = G \cdot a_{in}(t) \cdot \cos(\omega_{LO}t) + \mathrm{j} \cdot G \cdot a_{in}(t) \cdot \sin(\omega_{LO}t) + n_{out,I} + \mathrm{j}n_{out,Q} \tag{4.12}$$

With G the conversion gain of one mixer, $a_{in}(t)$ the input signal and $n(t)$ independent white noise sources at the output of the downconverters. The expected value for the total quadrature noise power is :

$$\begin{aligned}
\left|n^2_{out,\text{quadrature mixer}}(t)\right| &= \left|\left(n_{out,I}(t) + \mathrm{j}n_{out,Q}(t)\right)^2\right| \\
&= \left|n^2_{out,I}(t) + 2\mathrm{j}\cdot n_{out,I}(t)\cdot n_{out,Q}(t) - n^2_{out,Q}(t)\right| \\
&= \left|n^2_{out,I}(t)\right| + 0 + \left|n^2_{out,Q}(t)\right| \\
&= 2\cdot\left|n^2_{out,\text{mixer}}(t)\right| \\
&= 2\cdot G^2\cdot\left|n^2_{in,\text{mixer}}(t)\right| \qquad (4.13)
\end{aligned}$$

$n_{in,mixer}$ is the input referred noise of one of the two mixers. An equivalent white noise source at the input of the complete quadrature downconverter, would give the following total output noise :

$$\begin{aligned}
\left|n^2_{out,\text{quadrature mixer}}(t)\right| &= \left|G^2\cdot n^2_{in,\text{quadrature mixer}}\left(\cos(\omega_{LO}t) + \mathrm{j}\cdot\sin(\omega_{LO}t)\right)^2\right| \\
&= G^2\cdot\left|n^2_{in,\text{quadrature mixer}}(t)\right|\cdot 1 \\
&= G^2\cdot\left|n^2_{in,\text{quadraure mixer}}(t)\right| \qquad (4.14)
\end{aligned}$$

The input referred noise of the total quadrature downconverter is thus 2 times higher than the input referred noise of the mixers. This means its dynamic range DR is 2 times lower, while its power consumption is 2 times higher (there are two mixers). The efficiency of a quadrature downconversion mixer is therefore 4 times lower than the efficiency of a double balanced mixer.

The Noise Efficiency η_n

η_n is the efficiency degradation due to excess noise sources. This accounts for all noise generated by devices different from the main dissipating elements which perform the wanted operation. In an OTA type of device η_n represents all noise generated by transistors different from the input transistors (in an OTA the input transistors generate the intrinsic noise of the transconductance). Fig. 4.7 shows an example of an OTA and indicates the main contributors to the excess noise for this OTA. In order to compensate for this excess noise, the power consumption will have to be increased. The magnitude of this efficiency degradation will mainly be determined by the design of the input stage of the building block.

The Power Efficiency η_p

η_p stands for the power efficiency with which the maximum output signal S is generated. η_p depends in first-order on whether the drivers are class A, class AB or class C (the maximal achievable efficiency for these stages is respectively about 30 %, 50 %

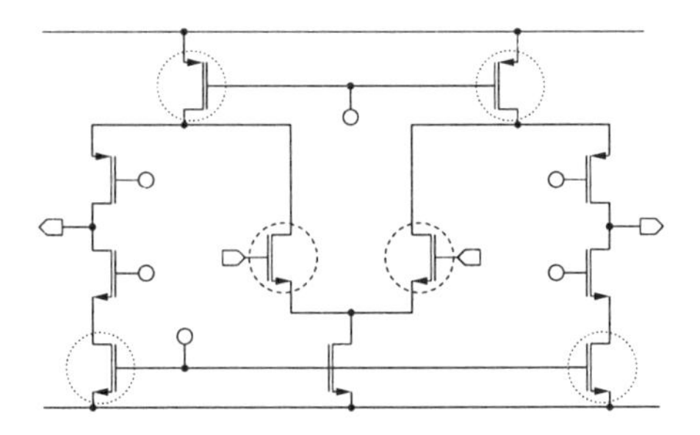

Figure 4.7. A folded cascode OTA and its main noise sources. The intrinsic noise sources are circled with a dashed line and the excess noise sources are circled with a dotted line.

and 70 %). A second factor in η_p is the power needed to drive parasitic capacitances. This makes that from a certain operation frequency on, there will be no linear relationship anymore between the power consumption of the building block and its working frequency (specified via the bandwidth BW). At higher frequencies this relationship will become quadratic or worse and the power consumption will start to increase rapidly in function of its working frequency. For bandpass filters this is even worse, because of the high ratio between the operating frequency F and the bandwidth BW. The intrinsic power consumption is determined by the bandwidth BW, but the increase due to parasitic capacitances is determined by the actual working frequency of the building block, being the highest of the bandwidth BW and the operating frequency F. The chosen implementation technique has also a large influence on the frequency dependence of the efficiency. The active-RC technique has for instance a better efficiency than the OTA-C technique at low frequency (higher signal swing, lower excess noise), but it is more sensitive to parasitic capacitances and therefore its efficiency decreases in function of frequency much faster than the efficiency of the OTA-C technique. This explains why the active-RC technique is better for low frequency implementations, while the OTA-C technique is better for high frequency applications. With the further improvement of technologies the trade-off point between the two shifts to ever higher frequency (due to smaller parasitic capacitances). For a typical 0.7 μm CMOS process this is for instance situated around a bandwidth BW of 10 MHz (or an operating frequency F of 10 MHz, if this is higher). Similar comparisons can be made between other implementation technique.

The Signal Efficiency η_s

η_s is the efficiency due to the limited output swing.

$$\eta_s = \left(\frac{S}{V_{DD}}\right)^2 \tag{4.15}$$

The power efficiency η_p described the efficiency with which a signal with swing S is generated. It does not make any assumption concerning the power supply V_{DD}. The largest signal that may be applied will be lower than what theoretically can be expected for a given output stage and a given power supply level (V_{DD} for class A and class AB, and $2V_{DD}$ for a class C output stage). η_s brings this effect into account. The maximum signal swing S that is allowed, is the signal level above which the distortion components become too high. 'Too high' is here defined as : when the magnitude of the distortion components is higher than the total integrated noise. Under these conditions the dynamic range DR is called 'distortion free dynamic range' [Groen CAS91]. Depending on the type of signal (digital data is for instance less sensitive to distortion than to noise), a relaxed specification which allows for easier calculations is sometimes used. A typical example is taking S at the point were the distortion is equal to -40 dB. The maximum allowable signal swing S is often a fraction of the power supply voltage ranging from about a fifth to a value almost equal to 1 (the latter is called rail-to-rail operation). Its value mainly depends on the chosen implementation technique. OTA-C filters will for instance have a lower η_s than active-RC filters. The signal efficiency η_s depends also on the chosen power supply voltage. It improves when higher power supplies are used. It is assumed that the maximum allowable input and output swing of a building block are equal. If not, their ratio must also be brought into account in the high-level design.

4.3 STRUCTURED DESIGN OF TRANSCEIVERS

4.3.1 *Trade-offs in Receiver Design*

The input signal of a receiver is the antenna signal. It is a broadband signal (several hundreds of MHz's) that consists of the wanted signal surrounded by several, larger and smaller, neighbor signals. In successive stages of filtering, amplification and downconversion, the unwanted neighbor signals are further and further suppressed and the wanted signal is brought down to lower and lower frequencies until the final, low dynamic range, low frequency signal can be sampled with a low or medium accuracy A/D-converter. The overall amplification that is realized in the receiver path from the antenna to the A/D-converter has to be distributed over the different stages. The value of this overall gain can vary widely depending on the signal strength of the incoming wanted signal. Its distribution over the different stages can also vary widely. It depends on the spectral distribution of the unwanted neighbor signals and their signal level compared to the wanted signal. The distribution of the overall gain over the different stages has to be done such that the maximum possible gain is realized in each stage (for optimum performance) while at the same time none of the stages is saturated under any circumstance. The more filtering is done in the early stages (with very high Q filters) the more the gain can be applied in these stages, resulting in a lower power consumption for the succeeding stages. The trade-off is that the cost of very high Q filters is high. It is not possible to do all the filtering and downconversion in one stage. This would impose unrealistic specifications on both the high frequency filter and the downconverter. The bandwidth of the wanted signal and its neighbors has to be reduced gradually as its center frequency is decreased. Performing accurate filtering at high frequencies comes at the cost of a high power consumption and a low degree of integratability. On the other hand, the more signal is filtered at high frequencies, the more relaxed the specifications become for the succeeding stages, resulting on its turn in a lower power consumption and a better integratability. This is the trade-off that has to be made in receiver design.

The classically used technique to find an optimum for these high-level trade-offs is experience driven and based on the trade-off description as given above. A structured design approach requires however a formal description of the high-level design problem. In the next section a technique for the formal description of the high-level receiver design problem is proposed and used to develop calculation techniques for the high-level optimization of a given receiver architecture for a given application. In this way this structured design approach allows for design automation and global optimization, giving faster and better results than experience driven design.

4.3.2 Trade-offs in Transmitter Design

There is a significant difference between the design of receivers and transmitters. The input signal to the transmitter is a low frequency signal which contains only the information which has to be modulated and upconverted to its carrier. This signal is well known in shape, amplitude and power distribution. This is completely different from the situation of a receiver. The input signal for a receiver is the antenna signal which is a highly random signal that can vary widely in amplitude and power distribution. The known component of the antenna signal, the wanted signal, is in some cases only a very small fraction of the total signal. The input signal of the transmitter is very well known and its amplitude and spectral power distribution are constant in time. There are therefore no trade-offs to be made during the high-level design of a transmitter. Once the behavioral models are known for the building blocks of the transmitter, high-level design is a straightforward calculation of the required building block specification starting from the input signal and the specifications set by the application.

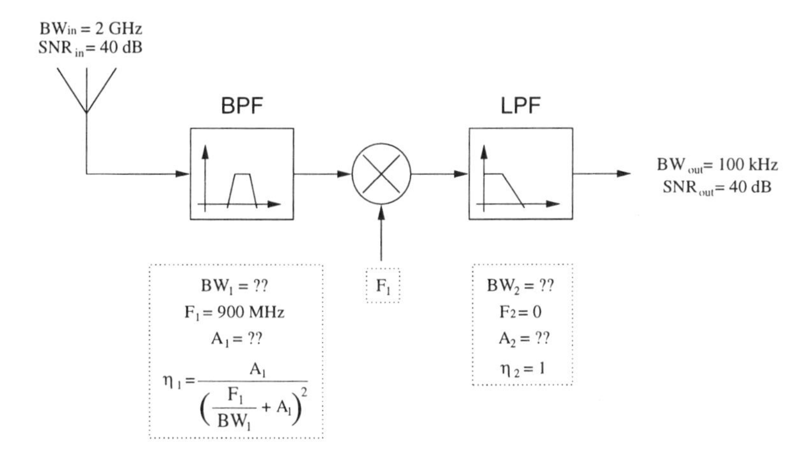

Figure 4.8. A simplified model of a receiver optimization problem with a bandpass filter, a lowpass filter and a perfect downconverter.

4.4 A DESIGN METHODOLOGY FOR RECEIVER ARCHITECTURES

4.4.1 An Example of Receiver Optimization

Fig. 4.8 gives a simplified example of the optimization of a receiver. It shows two filters, a bandpass filter and a lowpass filter. In this analysis both filters are assumed to be brick wall filters (i.e. they have a flat passband and they totally suppress all out of band signals). The downconversion mixer between the two filters is assumed to be perfect. A broadband input signal, for this analysis assumed to be white ($SNR_{in} = SNR_{out}$), has to be filtered and amplified. The allowable signal swing SW is assumed to be equal for all nodes.

The amplitude of a signal S (its rms value) is reduced when it is filtered in a building block. The gain that may be implemented in this building block is equal to this reduction. More gain is not allowed because this would saturate the output stage of the building block, while the use of less gain is not advisable because it only makes the noise specifications for the succeeding building blocks higher. The signal reduction without gain can be found by integrating the PSD ($s^2(f)$) of the signals :

$$\frac{S_{out}^2}{S_{in}^2} = \frac{\int s_{out}^2(f) \cdot \mathrm{d}f}{\int s_{in}^2(f) \cdot \mathrm{d}f} \tag{4.16}$$

In this example the input signal is a white signal. Its PSD is therefore constant within its passband.

$$\frac{S_{out}^2}{S_{in}^2} = \frac{s^2 \cdot \int_{BW_{out}} \mathrm{d}f}{s^2 \cdot \int_{BW_{in}} \mathrm{d}f} = \frac{s^2 \cdot BW_{out}}{s^2 \cdot BW_{in}} = \frac{BW_{out}}{BW_{in}} \tag{4.17}$$

A is the overall gain of the total receive path. The gain that can be applied in a building block is equal to the signal reduction.

$$A = \frac{S_{in}}{S_{out}} = \sqrt{\frac{BW_{in}}{BW_{out}}} \tag{4.18}$$

This gain has to be divided over the two stages.

$$A = A_1 \cdot A_2 \tag{4.19}$$

By setting these two gains, the bandwidths BW_i and dynamic ranges DR_i of the two building blocks will also be determined. For the gain of the first stage A_1 the same equations as for the overall gain A holds :

$$A = \sqrt{\frac{BW_{in}}{BW_1}} \Rightarrow BW_1 = \frac{BW_{in}}{A_1^2} \tag{4.20}$$

The overall output bandwidth BW_{out} is realized with the last stage :

$$BW_2 = BW_{out} \tag{4.21}$$

The dynamic range of a building block is defined as the maximal allowable input signal over its input referred noise (N is the integrated noise, n^2 is the noise density)

$$\begin{aligned} DR_1 &= \frac{S_{in}^2}{N_1^2} = \frac{\int_{BW_{in}} s_{in}^2(f) \cdot \mathrm{d}f}{\int_{BW_1} n_{in}^2(f) \cdot \mathrm{d}f} \\ &= \frac{s^2 \cdot BW_{out}}{n^2 \cdot BW_1} = SNR \cdot \frac{BW_{in}}{BW_1} = SNR \cdot A_1^2 \end{aligned} \tag{4.22}$$

The same goes for the second filter :

$$DR_2 = SNR \cdot A_2^2 = SNR \cdot \frac{A^2}{A_1^2} \tag{4.23}$$

The division of the overall gain A over the two stages has to be done in an optimal way. In this example a minimum power consumption is required. P follows from equation 4.3 :

$$P = \eta_1 \cdot DR_1 \cdot kT \cdot BW_1 + \eta_2 \cdot DR_2 \cdot kT \cdot BW_2 \tag{4.24}$$

The efficiency of a lowpass filter is 1. For an active integrated bandpass filter the efficiency is (see section 4.2.2.2) :

$$\eta_1 = \frac{A_1^2}{(1 + 2Q + A_1)^2} \approx \frac{A_1^2}{4Q^2} = \frac{A_1^2 \cdot F_1^2}{4 \cdot BW_1^2} \tag{4.25}$$

The power consumption can be calculated in function of A_1 and the known parameters.

$$P = \frac{1}{4} \cdot kT \cdot SNR_{in} \cdot \frac{F_1^2 \cdot A_1^6}{BW_{in}} + kT \cdot SNR_{in} \cdot \frac{BW_{in}}{A_1^2} \tag{4.26}$$

This function of A_1 has a minimum for :

$$A_1 = \sqrt[8]{\frac{4}{3} \cdot \frac{BW_{in}^2}{F_1^2}} \tag{4.27}$$

For this example the optimum value for A_1 is thus very small (1.27) and BW_1, the bandwidth of the first stage, is large (1.2 GHz). This result could be expected because of the low efficiency of high Q (i.e. small bandwidth) active integrated bandpass filters.

4.4.2 General Formulation of the Optimization Problem

In the previous section a lot of simplifications and approximations were made in order to make the optimization problem calculatable by hand. Only two building blocks were considered, only one parameter was allowed to vary freely and signals were assumed to be white and invariable, excluding also the need for functions with variable gain. The general receiver optimization problem is much more complex. A high-level receiver design optimization problem for a given application can be described in the following way : for a given application and a given receiver architecture, find for each building block the specifications which will render the minimum overall power and/or area consumption power consumption and which will meet the required signal reception quality under all specified conditions.

A formal description of this problem is necessary in order to be able to find a formal solution of the high-level design problem. This formal description is given here. A receiver is specified as an ordered set of building blocks BB_i :

$$(BB_i) = ([T_i \; F_i \; BW_i \; A_i \; DR_i]) \tag{4.28}$$

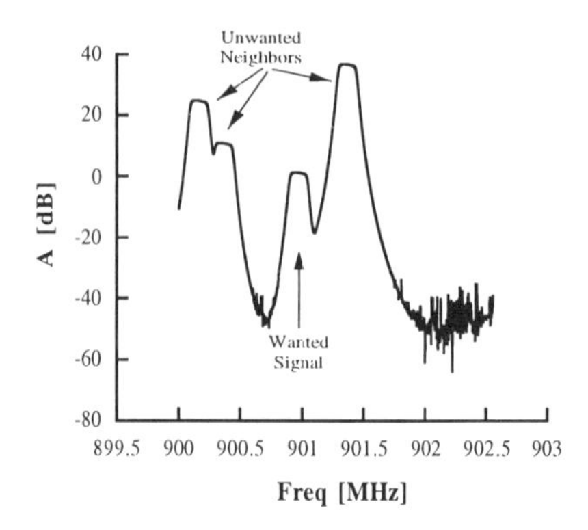

Figure 4.9. Example of an input spectrum.

The receiver architecture is defined by the ordered set of building block types (T_i). The receiver architecture is given as input for the high-level design problem. The building block specifications must be found as result of the high-level design.

Fig. 4.9 gives an example of the spectrum of a possible antenna signal for a typical application. This signal is defined as S_{0j} (it is the j-th possible input spectrum of the first receiver building block; more general is S_{ij} the output signal of the i-th building block when S_{0j} is applied). An applications is described with an unordered set of possible input spectra $\{S_{0j}\}$. The receiver must be able to handle each of these possible input conditions. The wanted signal occupies only a small part of the input signal spectrum (in fig. 4.9 it is situated at 900.95 MHz $\pm$ 100 kHz) and its amplitude, its center frequency and its amplitude relative to its neighbors, can vary widely for the different possible input conditions. The overall gain of the complete receiver path and the distribution of this gain (and thus the filtering) over the different stages of the receiver are both highly dependent on the spectral distribution of the applied antenna signal. Every filtering operation in each building block will allow, for a certain antenna spectrum, a certain amplification.

A_{ij} is defined as the gain that can be allowed for building block BB_i under input spectrum S_0j. In a non-AGC building block the gain A_i of the building block is fixed and, in order to prevent saturation under all possible conditions, it must be equal to the smallest A_{ij} value.

$$A_i = \min_j(A_{ij}) \tag{4.29}$$

This has a direct influence on the dynamic range requirement of building block BB_i. When A_i is known, the relationship between the dynamic ranges and bandwidths of two consecutive stages BB_{i-1} and BB_i can be determined (SW is the maximum signal swing allowed on a node between two building blocks; for now, it is assumed to be equal for all nodes) :

$$\begin{aligned} DR_i &= \frac{SW^2}{N_i^2} = \frac{SW^2}{\int n_i^2 \cdot \mathrm{d}f} \approx \frac{SW^2}{n_i^2 \cdot BW_i} \\ &= \frac{SW^2}{A_{i-1}^2 \cdot n_{i-1}^2 \cdot BW_i} = \frac{SW^2}{A_{i-1}^2 \cdot n_{i-1}^2 \cdot BW_{i-1}} \cdot \frac{BW_{i-1}}{BW_i} \\ &= DR_{i-1} \cdot \frac{1}{A_{i-1}^2} \cdot \frac{BW_{i-1}}{BW_i} \end{aligned} \tag{4.30}$$

Small gains result in high dynamic range levels. A higher dynamic range is equivalent to lower noise level requirements and this can only be realized at the cost of a higher power and area consumption. It is thus important to find, for a given set of input spectra $\{S_0 j\}$, a set of bandwidths BW_i, gains A_i and center frequencies F_i which gives the minimal power and/or area consumption. The center frequencies are important because most building blocks will require more power at higher frequencies to realize a certain bandwidth/gain/dynamic-range specification.

The specifications imposed by the application can not only be defined via a set of unordered input spectra $\{S_0 j\}$. One aspect is that in some applications the required performance may vary with the reception conditions. A lower performance may for instance be allowed when a very low wanted signal is received in combination with very high unwanted signals. This means that for each different input spectrum, a different signal-to-unwanted-signal ratio $SUSR_j$ may be defined. Therefore, an application is here defined with an unordered set of input conditions $\{I_j\}$. An input condition I_j from this set contains an antenna input spectrum S_{0j} (as shown in fig. 4.9), the signal-to-unwanted-signal ratio $SUSR_j$ that is required for this given input spectrum and the probability PR_j with which such a situation can occur. The probability PR_j of an input condition is only necessary to calculate the power consumption of building blocks with an automatic gain control (AGC) function because their power consumption may vary under the different input conditions. Finally, for each input spectrum S_{0j} it must be indicated at which frequency RF_j the wanted signal is located. This is necessary for the performance evaluation of complex receiver architectures, like the wideband-IF receiver [Briant ACD96], in which the wanted signal is not downconverted to a fixed frequency. A set of input conditions is thus defined as :

$$\{I_j\} = \{[S_{0j}\ SUSR_j\ RF_j\ PR_j]\} \tag{4.31}$$

4.4.3 Calculation of the Receiver Performance

Fig. 4.10 proposes a flow diagram for the calculation of a receiver's performance. Here, this is called the forward calculation of the performance : the receiver topology and all building block specifications are given and it is only the performance that has to be determined. In the proposed calculation diagram each block converts, as was shown in fig. 4.1, an input signal $S_{(i-1)j}$ into an output signal S_{ij} via a frequency translation (FS), a filter operation (LTF) and an amplification (A) (in fig. 4.10 jointly called the 'Building Block Transfer Characteristic'). The difference between the proposed computation scheme and the computation scheme of fig. 4.1 is that the unwanted noise, distortion and aliasing signals are all calculated separately, in parallel to the wanted signal. It is only after the calculation of the total receive path, at the output of the receiver, that these unwanted signals are compared to the amplitude of the wanted signal in its passband. An advantage of the proposed computation scheme is that the results of all operations can be monitored on each intermediate node. Unwanted operations on unwanted signals generated in previous building blocks do not have to be calculated. The specified receiver will have either a correct operation point and all noise, distortion and aliasing components will be small so that unwanted operations on them will give negligible results, or the receiver will have an incorrect operation point and in that case the results of the performance analysis can only be used to determine that the proposed building block specifications must be rejected.

Three different types of effects must be calculated in order to be able to determine the performance of a receiver :

- the signal transfer from the antenna signal to each intermediate node;
- the frequency shifting and aliasing effects;
- the distortion and intermodulation.

The signal transfer can not be determined with the classical AC simulation. An AC simulation gives a transfer function, i.e. it gives the relationship between an input and output spectrum. This must be combined with the information of the power spectral density distribution in order to be able to determine the absolute signal levels on each node. A simulation technique which keeps track of both the power distribution on each node and the transfer functions between every two succeeding nodes is thus necessary.

A transient analysis could theoretically be used for this purpose. A data-sequence, similar to a sampled version of the actual antenna signal, could be applied at the input and the power spectral density distribution on each node could be determined by taking the FFT of the time domain signals. However, this would require very large sets of data-points (a 1 kHz resolution requires more than 1 million datapoints per signal), resulting in a massive memory and calculation time consumption. Another problem is the non-linear operations. It is not a problem to perform a nonlinear operation in a transient analysis, the problem is how to discriminate in the result between the wanted (frequency shift) and the unwanted (aliasing, distortion) non-linear operations.

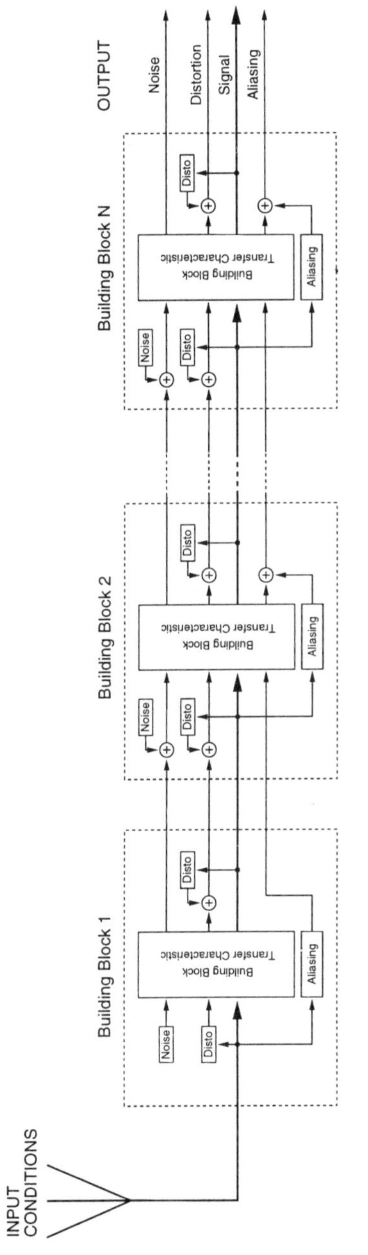

Figure 4.10. Signal flow for the proposed algorithm to calculate the performance of a receiver.

A method to get a good insight in the different types of nonlinear behavior in RF circuits is the harmonic balance simulation technique [Kund TCAD86]. The effects of distortion and aliasing can be observed quite accurately, but it is not practical with this technique to evaluate the actual overall performance reduction due to these effects. This would again require the use of more than 1 million data-points (now in the frequency domain) for the representation of the power spectral density distribution.

An important aspect of the implementation of the performance calculation is thus the representation of the signals S_{ij}. S_{ij} is a power spectral density ranging from DC to several GHz. The resolution has to be about 1 kHz, because the wanted signal is only a very small part (a few hundred kHz) of the total power spectral density. A data-point representation of the signal S_{ij} requires more than 1 million points, which is highly impractical. Here, an approach is proposed in which the signals S_{ij} are symbolically represented and all operations are implemented as formula manipulations. A signal S_{ij} (a power spectral density distribution) is represented in the simulator as a sum of rational polynomials RP_m, each restricted to a non-overlapping frequency interval :

$$S_{ij}(f) = \sum_m RP_{ijm}(f) \tag{4.32}$$

$$\text{with} \quad RP_m(f) = \begin{cases} \dfrac{n_{m0} + n_{m1}f + n_{m2}f^2 + \ldots + n_{mk}f^k}{d_{m0} + d_{m1}f + d_{m2}f^2 + \ldots + d_{ml}f^l}, \\ \qquad \forall f \in \left]f_{begin,m}, f_{end,m}\right], \\ 0, \quad \forall f \in \left]-\infty, f_{begin,m}\right] \cup \left]f_{end,m}, +\infty\right]. \end{cases}$$
$$\text{and} \quad \forall\, m \neq n : \left]f_{begin,m}, f_{end,m}\right] \cap \left]f_{begin,n}, f_{end,n}\right] = \varnothing$$

This representation technique gives a high flexibility and a very low memory consumption. Most power spectral density shapes are in fact of the rational polynomial form and those which are not, can be represented by fitting the actual shape in small intervals to rational polynomials with a limited number of coefficients. The use of the power spectral density instead of the amplitude and phase of the signals has the disadvantage that the phase information of the signals is lost. However, the analyzed receiver topologies do not have any loops or branches and the phase information is only important when two correlated signals join (in a loop or branch) so that they may fully cancel due to a 180° phase difference. The power spectral density representation is used because the different frequency components of the signal spectrum are either uncorrelated or, when very close together (closer than the correlation bandwidth, e.g. 200 kHz), correlated, but with an unknown phase relationship. For such signals the relationship holds that the power spectral density of the sum of two signals is equal to the sum of their power spectral densities. An effect that can not be analyzed due to the fact that the phase information is omitted, is the performance degradation due to a non constant group delay for the wanted signal. The group delay is the derivative of the phase and a non linearity of the phase can be compensated by using an equalization algorithm in the DSP after sampling.

The proposed formula manipulations are easy and fast : filtering is multiplying rational polynomials with rational polynomials, resulting again in a rational polynomial, and a frequency translation is done by replacing f with $f - F_i$ and calculating the new coefficients.

$$\begin{aligned}
\mathrm{LTF}\left(S_{ij}(f)\right) &= \left(\frac{a_0 + a_1 f + \ldots + a_k f^k}{b_0 + b_1 f + \ldots + b_l f^l}\right) \cdot S_{ij}(f) \\
&= \left(\frac{a_0 + a_1 f + \ldots + a_k f^k}{b_0 + b_1 f + \ldots + b_l f^l}\right) \cdot \sum_m RP_{ijm}(f) \\
&= \sum_m \left(\left(\frac{a_0 + a_1 f + \ldots + a_k f^k}{b_0 + b_1 f + \ldots + b_l f^l}\right) \cdot RP_{ijm}(f)\right) \\
&\doteq \sum_m RP_{LTF,ijm}(f)
\end{aligned} \tag{4.33}$$

$$\begin{aligned}
\mathrm{FT}\left(S_{ij}(f)\right) &= S_{ij}(f - F_i) \\
&= \sum_m RP_{ijm}(f - F_i) \\
&\doteq \sum_m RP_{FT,ijm}(f)
\end{aligned} \tag{4.34}$$

By taking into account both the even and odd terms in the power spectral density representation of equation 4.32, quadrature signals can also be represented as single signals. The odd terms make it possible to create and represent power spectral densities which are different for positive and negative frequencies, giving in this way the correct representation of a quadrature signal and quadrature operations like a quadrature downconversion. The distortion of a signal $S_{ij}(f)$ must of course be calculated starting from the two separate components of the quadrature signal and not from the combined representation. A reconstruction of the separate I and Q components from the quadrature representation is in this approach still possible at any time.

The noise power spectra $N_i(f)$ and the unwanted aliasing spectra $AL_{ij}(f)$ can also be represented with the formula of equation 4.32. Distortion components can however not be represented with equation 4.32. A distortion spectrum $D_{ij}(f)$ is proportional to the convolution of 2 (2nd order) or 3 (3rd order) signals $S_{ij}(f)$. A convolution requires an integration operation. The integral of a rational polynomial is in general not a rational polynomial anymore. Distortion components are therefore best handled and stored as a sum of terms that consist of the convolution of two or three rational polynomials

Distortion components of 4th and higher order can be neglected compared to 2nd and 3rd order distortion. By limiting the distortion analysis to second- and third-order distortion and intermodulation, and by not calculating distortion of distortion components, the distortion components can be represented in a closed form and a recursive definition is avoided.

$$\mathrm{D2}\left(S_{ij}(f)\right) = d_2 \cdot \left(\int S_{ij}(f') \cdot S_{ij}(f - f') \cdot \mathrm{d}f'\right) \tag{4.35}$$

$$\mathrm{D3}\left(S_{ij}(f)\right) = d3 \cdot \left(\int S_{ij}(f - f') \cdot \left(\int S_{ij}(f' - f'') \cdot S_{ij}(f'') \cdot \mathrm{d}f''\right) \cdot \mathrm{d}f'\right) \tag{4.36}$$

The distortion components do not have to be calculated over the entire frequency range, nor do they have to be evaluated at intermediate nodes in the receiver topology. Instead, formulas 4.35 and 4.36 can be stored in closed form and they only have to be evaluated at the output of the receiver in a small passband (the bandwidth of the wanted signal).

4.4.4 Calculation of the Gains

Fig. 4.11 gives a flow diagram for the backward calculation of the performance problem. Fig. 4.10 showed the forward calculation problem where the receiver and the input conditions are fully specified and the performance is calculated. In the backward calculation of the performance problem the required performance is specified and all receiver building blocks are also fully specified except for their gains A_i and their dynamic ranges DR_i. These gains and their corresponding dynamic ranges will be the result of the backward calculation.

First all unscaled ($A_i = 1$) signal levels US_{ij} are calculated with the computation scheme of the forward calculation problem for the wanted operations as given in the previous section. Starting from the first stage, A_i must then be calculated by making the highest possible signal on node i (the output of building block BB_i equal to the maximum allowable swing SW). signal swing

$$SW = \max_j(S_{ij}) = \left(\prod_1^i A_k\right) \cdot \max_j(US_{ij}) \tag{4.37}$$

This formula does not hold for stages with automatic gain control (AGC). For such stages two boundaries can be found between which the gain varies during operation. From A_i, BW_i and the required SNR at the output, the dynamic ranges DR_i can be found by back calculation from the last stage to the first stage. With the following formula all building blocks will have the same noise contribution in the final SNR.

$$DR_{i-1} = DR_i \cdot A_{i-1}^2 \cdot \frac{BW_i}{BW_{i-1}} \tag{4.38}$$

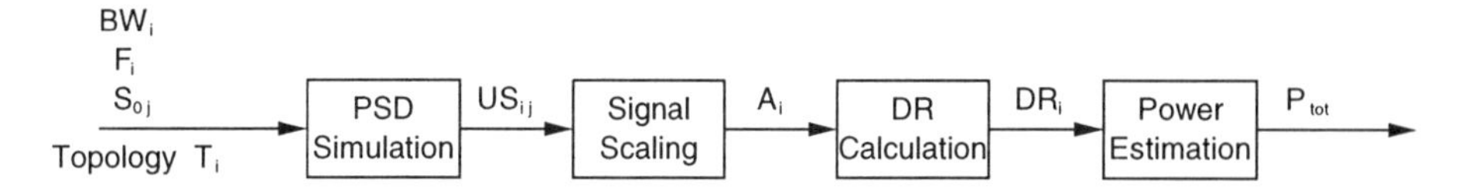

Figure 4.11. Calculation of the optimal receiver when its frequency behavior and the required performances are specified.

The receiver is fully specified once A_i and DR_i are known. The power consumption P_i of each stage and the overall power consumption P_{tot} can be estimated with the behavioral models of section 4.2.2.2.

With the presented backward calculation method the optimal receiver building block specifications are found when only the receiver topology, its frequency behavior (downconversion frequencies and filter bandwidths) and the specifications imposed by the application are known. In this way a receiver can be designed very fast when its frequency behavior is already fixed. The proposed technique has however also some disadvantages. The unwanted operations (noise, distortion and aliasing, as shown in fig. 4.10) are in the proposed computation scheme for the backward calculation problem not calculated and their effects are only partially taken into account. Aliasing effects, especially mirror signals, for instance are not taken into account, but it may be assumed that these effects have been examined before because the frequency behavior is already fixed. With equation 4.38 the noise of each building block is determined so that all building blocks generate the same amount of noise when referred to the input. With this amount the required performance specification is met only when all other building blocks would not generate any noise. Obviously, this means that the total accumulated input refered noise level would be much too high. Therefore, an extra noise margin will have to be added to all building block specifications, weighed with their power consumption. To building blocks with a low power consumption a high noise margin can be added, while building blocks with a large power consumption can accept only a limited noise margin. Starting from the backward calculation, good noise levels for all building blocks can be found easily. The distortion of the building blocks has been taken into account via their input and output signal swing SW. These are, depending on the building block type and the applications sensitivity to distortion, set for instance to result in a distortion of 1 %. A distortion free dynamic range specification can not be implemented with the backward calculation method, because this assumes a dependence of the signals swings SW_i on the dynamic ranges DR_i which still have to be calculated.

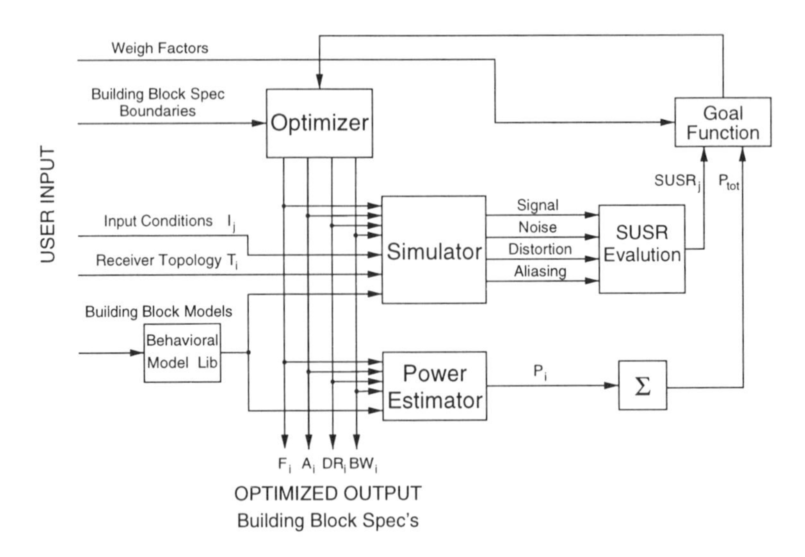

Figure 4.12. Schematic representation of the proposed optimization algorithm.

4.4.5 Optimization and Experimental Results

The chosen frequency behavior of a receiver, specified via the F_i and BW_i of the building blocks, has a large influence on the achievable performance and in a proper optimization all building block parameters (F_i, BW_i, A_i and DR_i) should therefore be varied freely (within specified boundaries). Fig. 4.12 gives a schematic diagram of such a full optimization procedure. The optimization algorithm searches for the set of building block parameters which satisfies the set of input conditions at the lowest power consumption. The performance of a given topology $\{T_i\}$ for an application specified via the input condition set $\{I_j\}$ is evaluated for a certain set of specifications with the simulation method based on formula manipulation of rational polynomials, as presented in section 4.4.3. A behavioral model library contains high-level models for the different types of building blocks. These models describe on one hand their wanted and unwanted operations (as described in fig. 4.10) and on the other hand it gives the relationship between the building block specifications (F_i, BW_i, A_i and DR_i) and their area consumption for a given power supply voltage V_{DD}. An optimization algorithm, such as simulated annealing [Laarh 1987], performs the optimization. The goal function for this optimization has $j + 1$ inputs. The goal function must be set such that the performance of $SUSR_j$ under input condition I_j meets the required performance while the total power consumption is minimized.

The methodology described in fig. 4.12 has been experimentally implemented as a computer automated tool, named ORCA (Optimizer for ReCeiver Architectures)

[Crols ICCAD95]. This experimental version of the tool uses Matlab as computation engine [Mathw 1992], interfaced to the main C++ program which implements a simulated annealing algorithm for the optimization loop.

ORCA can be used interactively, as a simulator, or as an automatic optimization tool. The combined IF zero-IF receiver topology of fig. 4.2 used to generate an example for both cases. When ORCA is used as a simulator, all building block specifications (BW_i, A_i, F_i, DR_i) are provided by the user. In this mode the RF designer can quickly compare different topologies or evaluate the influence of different building block specifications on the overall performance of the receiver. The simulator calculates symbolically the different unwanted signals (noise, aliased signals and distortion) on the different nodes. An actual numerical evalution is of the signal-to-unwanted-signal ratio is only performed on the output node and only in the signal band of interest (for this example a 100 kHz band at baseband). In this way a fast and memory efficient implementation was obtained. Simulation times go from less than 1 minute to several minutes (on a SUN Sparc10) depending on the number of different input conditions that are specified by the user. The required memory size for data storage is about 1 kB.

ORCA can also be used in an automatic mode, as an optimization tool. In that case the RF designer has to specify the optimization variables (which building block specs have to be varied during optimization, e.g. de bandwidth BW of the bandpass filter BPF), the boundaries for each optimization variable (e.g. 20 - 800 MHz for BW_{BPF}) and the cost function (a weighted sum of the total power and the deviation from the $SUSR$ specification). Due to the nature of simulated annealing, one optimization run takes typically several hours. It can however be expected that a full C++ implementation would significantly improve this value. Table 4.1 shows the results obtained from an optimization run for a specified SUSR of 40 dB. The total power required by the receiver was 50 mW. All italic numbers are optimization results. SE_i is the input signal sensitivity in a 200 kHz passband (defined as $DR_i \cdot BW_i/200$ kHz). Fig. 4.13 shows a power versus $SUSR$ plot obtained from several runs with different $SUSR$ specifications.

4.5 CONCLUSION

The need for a further integration of the analog transceiver front-end is bringing an evolution into the world of RF design. The introduction of new architectures makes it possible to come to highly integrated solutions, giving big advantages concerning cost and power consumption reduction. The evaluation of the qualities of the new architectures for a specific application is a problem. Estimating the power and area consumption and calculating the building block specifications for a given architectures and application has until now been an experience-based process, which implies only a limited form of optimization. For new architectures this is not possible anymore and a

Table 4.1. Optimization results for the combined IF zero-IF receiver topology with a SUSR requirement of 40 dB.

T_i	F_i [MHz]	BW_i [MHz]	A_i [dB]	DR_i [dB]	SE_i [dB]	P_i [dB]
BPF	910.0	54.0	0.0	∞	∞	0.0
LNA	0.0	2680.0	28.0	50.0	119.0	30.5
MIX 1	793.0	1738.0	3.4	55.0	98.0	14.5
SAW	117.0	55.0	0.0	∞	∞	0.0013
IF AMP	0.0	410.0	14.0	53.0	100.0	2.8
MIX 2	117.0	384.0	14.0	45.0	92.0	1.1
LPF	0.0	1.23	1.2	74.0	83.0	0.126
ADC	0.0	8.61	0.0	65.0	81.0	0.216

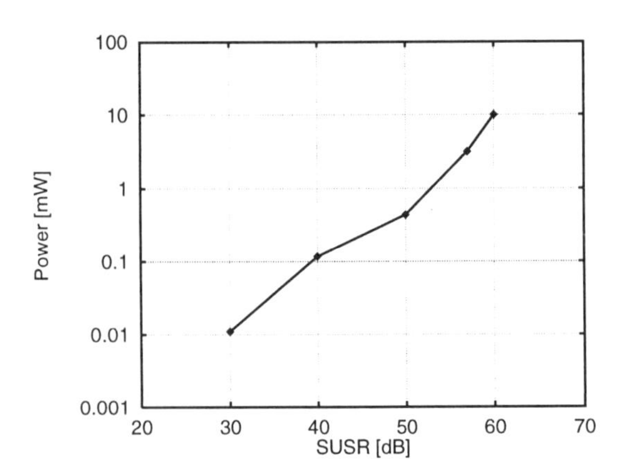

Figure 4.13. Optimal power consumption versus SUSR.

high-level design technique which allows for a fast and accurate calculation and optimization of the building blocks specification and an estimation of the power and area consumption for any given architecture and application is therefore necessary. In this chapter such a generally applicable high-level design technique has been proposed. It consists of three essential parts :

- The behavioral modeling of the different types of building blocks. The characterization of their wanted and unwanted signal operations and the modeling of the relationship between the building block specifications and its cost parameters like power and area consumption.
- The introduction of a new simulation method used to determine all wanted and unwanted signals (noise, distortion, mirror frequencies, self-mixing, ...). The new method is based on the representation of power spectral densities as rational polynomials and convoluted rational polynomials. Signal operations are implemented as formula manipulations. In this way it allows for high accuracy simulations with a low memory and calculation time consumption (a typical simulation time of 1 minute at a memory cost of about 1 kB). The results of the proposed calculation technique are used to determine the performance of a transceiver front-end.
- A backward calculation technique and an optimization algorithm are proposed for the high-level optimization of receiver architectures for a given application. The proposed techniques translate high-level specifications, set by the application, in the most optimum building block specification that results in the lowest overall power and area consumption.

For each part of the high-level design, i.e. the behavioral modeling, simulation and optimization, a formal description has been given which can be used to accurately describe any high-level design problem for any type of application, architecture and building block. By proving that a formal description of the high-level design problem is possible, the proposed techniques open the way to a computer automated implementation. An actual experimental implementation of the proposed high-level RF design methodologies as computer-aided design tool was presented [Crols ICCAD95] and the results of the automated high-level design of a typical receiver architecuture were discussed.

5 HIGH-LEVEL SYNTHESIS

5.1 INTRODUCTION

In chapter 4 the theoretical principles of high-level design have been introduced. An actual high-level design for a given application and a given architecture is given in this chapter with the GSM system taken as demonstrator. The complete high-level design of a highly integrated transceiver front-end will be performed for a mobile GSM handset. All the new techniques presented in chapters 3 and 4 will be used to propose a topology that achieves a level of integration and performance that goes far beyond present day state-of-the-art realizations. Starting from the specification for the GSM system and an architecture selection, the most optimimal building block specifications for the chosen architecture are derived.

Although a formal method for high-level design and optimization has been proposed in chapter 4, this method will not be used in the presented practical design example to its full extend. This design example is set to comply with the GSM standard as defined by the ETSI (European Telecommunications Standard Institute). This standard is given, more or less, as a number of situations under which each time a certain performance must be achieved. This is the same representation as is used in the formal high-level design method proposed in the previous chapter. The problem with standards for present day applications is that they are simplified in such a way that they allow for hand calculation and experience driven high-level design, rather than structured high-level design. The number of possible situations and required performances is drastically reduced, resulting in an over-specification and a reduced possibility for optimization. The formal high-level design and optimization allows for a structured high-level design approach and a much better optimization and power consumption reduction, when a more elaborate description of the practical requirements for the GSM system would be available. The results given in this chapter are based on this structured high-level design methodology, but the nature of the GSM standard brings them close to an experience based high-level design.

5.2 DIGITAL WIRELESS APPLICATIONS

Today, many new digital wireless applications are being introduced. Some have to replace and improve existing, analog, applications, others are new applications for new markets. The following list gives only a brief and limited overview of some of these systems and their application :

- **GSM :** GSM (Global System Mobile) is the European digital wireless system for person-to-person voice and data communication around 900 MHz. It uses typically medium sized cells (radii of approximately 10 km) and a 2 Watt transmission power. Originally developed in Europe, it is now also very successful in Asia, Africa and South-America.

- **DECT :** The DECT system (Digital European Cordless Telephone) uses smaller cells than GSM for indoor and office use (radii of approximately 100 m) [McDon CICC92]. Its operating frequencies are situated at 1880 MHz.

- **NADC :** NADC stands for North American Digital Cellular. It is a North American alternative for the GSM system. Different from GSM, it uses spread spectrum techniques for channel multiplexing.

- **DCS 1800 and DCS 1900 :** DCS 1800 and DCS 1900 are almost exact copies of the GSM system, implemented however at respectively 1800 MHz and 1900 MHz. They are used in countries where either the GSM band is already occupied by another application or where more bandwidth is needed than available in the GSM band.

- **ERMES :** ERMES is the European low cost system for paging applications (European Radio Messaging System). Base stations cover areas with radii of up to 100 km. Its operating frequencies are located at 170 MHz.

- **ISM :** The ISM band (Industrial-Scientific-Medial) is a North American frequency band reserved for small range unlicensed wireless communication [Hull ISSCC96].

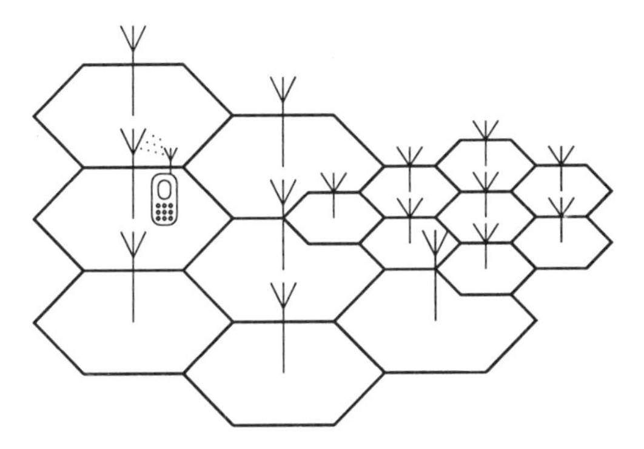

Figure 5.1. GSM is a cell based system, a mobile station communicates with the nearest base station.

Applications in this band use the spread spectrum technique to avoid interference between different users and different types of applications. The ISM band is situated around 880 MHz.

A more elaborate introduction to the many different digital wireless systems that exist can be found in [Rapal 1994].

5.3 GSM

5.3.1 The GSM System

Among the different wireless communication systems presented in the previous section, the GSM system has played during the passed years the role of technology driver. It was the first wireless system based on a digital and cell based technology that was introduced in a mass market. Many new techniques were therefore first developed for the GSM system. In this text, the GSM system has therefore been chosen as the demonstration application for which the proposed techniques of chapters 3 and 4 are used to do the high level design of a fully integrated transceiver chip in an optimal way. The GSM system is a system with a wide spread use and a typical configuration. It is a cell based system with mobile stations communicating with a network of base stations (see fig. 5.1). It is a band limited system which has its own bands of operation (890 MHz to 915 MHz for mobile transmission and 935 MHz to 960 MHz for mobile reception) wherein no signal foreign to the GSM system may be transmitted. Its operation bands are subdivided into different communication channels with 200 kHz spacing. It uses a TDMA (Time Domain Multiple Access) technique in which transmission and reception of the mobile station at the same time is not possible. Fig. 5.2 shows a frame sequence in the time domain for the GSM system.

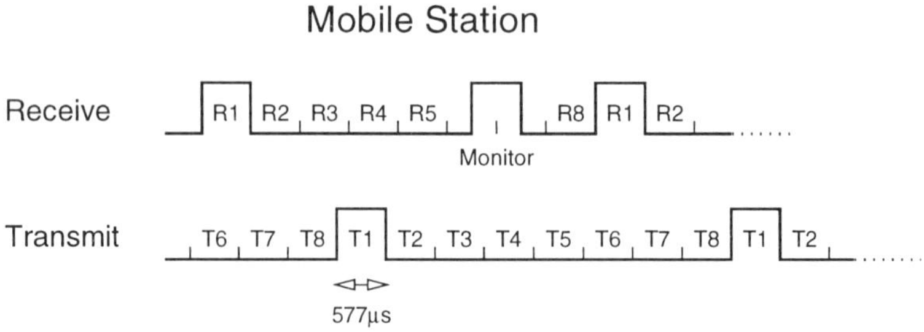

Figure 5.2. Typical sequence of frames within one receive and one transmit channel, shown in the time domain.

The GSM system has a setup which is, more or less, similar to most digital wireless systems. Compared to other applications it has very high specifications concerning low input noise, required dynamic range and required synthesizer accuracy. This makes the GSM system a good example of a high performance digital wireless application. The high-level design given in this chapter will be done for the GSM system, but most aspects can be used for the high-level design of a transceiver for any other wireless application.

The following sections give an overview of the GSM specifications as given by the ETSI (European Telecommunications Standard Institute) in [ETSI94], as far as they are important to the high level design of the transceiver and as far as they apply to the mobile unit.

5.3.2 GSM Receiver Specifications

5.3.2.1 Sensitivity. The required performance for a mobile station GSM receiver is specified in function of the bit error rate (BER), the frame error rate (FER) and the residual bit error rate (RBER, i.e. the bit error rate within a frame classified as 'good') which must be met for different types of channels and propagation conditions. BER, FER and RBER are however specifications which are very unpractical for designing. For a certain design, their values can not be obtained from calculations. They can only be obtained with lengthy BER simulations on a dedicated simulator with special libraries which model these channels and propagation conditions. BER, FER and RBER are a good way to specify the required performance of the GSM system as it takes into account all possible effects (like e.g. channel fading) and looks at the specification which is overall important : the amount of received useful information. This representation is however not necessary for the design of the analog part of the mobile receiver. The mobile receiver has, as a part of the total GSM system, no influence on many of these effects. Fading is for instance generated in the communication channel and corrected with an equalizer in the digital signal processor (DSP) after reception. The receiver can only deteriorate the performance by adding unwanted signals (like noise, mirror signals and intermodulation signals) to the wanted signal. The reference sensitivity performance (this is the performance which must at least be met under all specified conditions) can therefore be translated in a signal-to-unwanted-signal ratio (SUSR) specification which must be met by the receiver at all times (see also section 4.1). An approximate value for this SUSR can be found by means of BER simulations. For the GSM system the required SUSR that meets the BER, FER and RBER specification is 9 dB [Fenk ACD95]. This value can be used for the design of the receiver, to make hand calculations and to do fast simulations. The value can only be an approximation because it is assumed that the performance will be deteriorated by the different types of unwanted signals in the same way. This has, in general, not to be true. The wanted signal may for instance be less sensitive to distortion (a SUSR of 5 dB may be allowed) than to noise (where the 9 dB SUSR must be met). These differences are however small and the 9 dB SUSR specification can therefore be assumed, while the actual performance can be checked after design with BER simulations.

The reference sensitivity level is defined as the lowest possible input signal level for which the reference sensitivity performance is still met when no other interfering signal is present. The reference sensitivity level is required to be -102 dBm [ETSI(6.2)]. This means that the total equivalent input unwanted signal level at the frequencies of the wanted signal is -111 dBm. Noise is the only source of signal degradation when no other interfering signal is present. The total equivalent input noise level must therefore at least be lower than -111 dBm.

5.3.2.2 Adjacent Channels. The reference interference performance is, like the reference sensitivity performance, also specified via BER, FER and RBER specifications, but again an equivalent SUSR of 9 dB is assumed as specification. The reference interference performance has to be met when one of the following random modulated signals is present together with the wanted signal [ETSI(6.3)] :

- an interfering signal in the same channel as the wanted signal (called 'co-channel interferer') 9 dB below the wanted signal level;
- an interfering signal in the channel directly adjacent to the channel of the wanted signal (at +200 kHz or -200 kHz) 9 dB above the wanted signal level;
- an interfering signal in the adjacent channel at ±400 kHz, 41 dB above the wanted signal level;
- an interfering signal in the adjacent channel at ±600 kHz, 49 dB above the wanted signal level.

Although the achieved performance due to these interfering signals will in first order be rather independent of the actual signal level of both the wanted signal and the

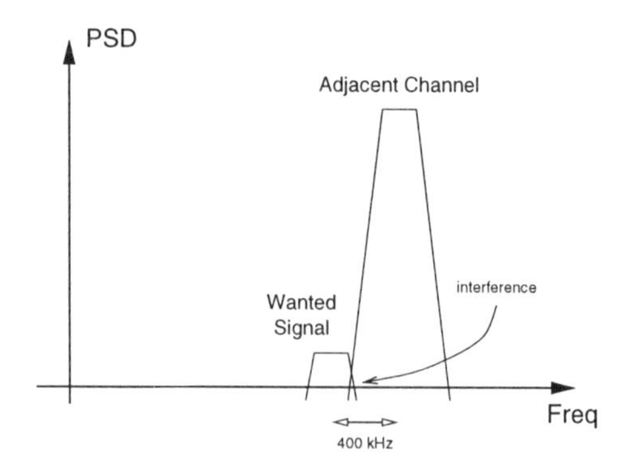

Figure 5.3. Degradation of the receiver performance due to a signal in the adjacent channel 400 kHz from the wanted signal.

Table 5.1. The in-band and the out-of-band blocking levels.

FREQUENCY					BLOCKING LEVEL
		IN-BAND			
600 kHz	$\leqslant$	$\mid f - f_0 \mid$	$<$	1.6 MHz	-43 dBm
1.6 MHz	$\leqslant$	$\mid f - f_0 \mid$	$<$	3 MHz	-33 dBm
3 MHz	$\leqslant$	$\mid f - f_0 \mid$			-23 dBm
		OUT-OF-BAND			
100 kHz	$\leqslant$	f	$<$	915 MHz	0 dBm
980 MHz	$\leqslant$	f	$<$	12.75 GHz	0 dBm

interfering signal, the interference ratios are specified for a wanted signal level of -85 dBm. Fig. 5.3 gives an example of an interfering signal at 400 kHz.

5.3.2.3 Blocking Signals. The effects of interfering signals further than 600 kHz away from the wanted signal are specified with the blocking signal characteristics [ETSI(5.1)]. This specification says that the reference sensitivity performance (9 dB SUSR) shall be met for a wanted signal 3 dB above the reference sensitivity level (i.e. -99 dBm) when a second signal, a continuous sine wave signal, is present at the level given in table 5.1.

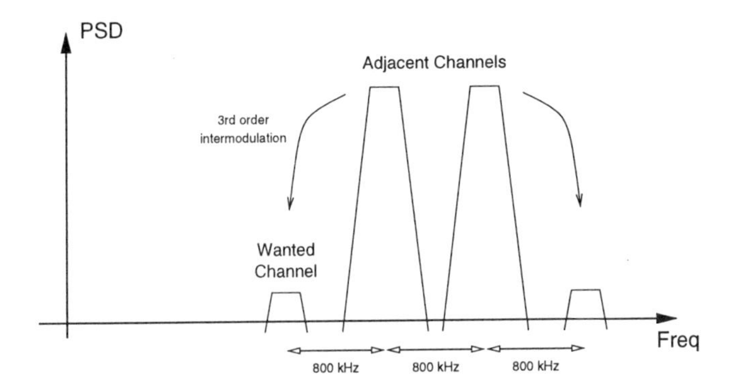

Figure 5.4. Degradation of the receiver performance due to intermodulation.

5.3.2.4 Intermodulation. A performance degradation will also occur when two signals generate a 3rd order intermodulation signal that lies at the wanted signal frequency. The reference sensitivity performance (9 dB SUSR) must also be met when the following signals are present [ETSI(5.2)] :

- a wanted signal of -99 dBm at a frequency f_0;
- a pseudo-random modulated signal of -43 dBm at a frequency f_1;
- a continuous sine wave signal of -43 dBm at a frequency f_2.

f_1 and f_2 must be placed at 800 kHz from each other and their 3rd order intermodulation component $(2f_2 - f_1)$ must be at the wanted signal frequency f_0. Fig. 5.4 shows this intermodulation effect in the frequency domain.

5.3.3 GSM Transmitter Specifications

5.3.3.1 Output Power. There are five different classes of mobile stations defined for GSM, depending on their maximum peak power. In this chapter only the class 4 mobile stations are considered. They have a 2 Watt maximum peak output power with a ± 2 dB tolerance [ETSI(4.1)]. This 2 Watt power class is today the most commonly used GSM mobile station for hand carried applications.

The nominal peak power level needs to be adaptive in steps of 2 dB for power level control and up and down ramping, starting from a 12 dBm level up to the maximum peak power level of 36 dBm. The power control steps have to form a monotone sequence and the required tolerance on the 2 dB intervals is ± 1.5 dB.

5.3.3.2 Output RF Spectrum. The output RF spectrum of the transmitted signal results from two effects :

Table 5.2. Specifications for the allowed RF output spectrum due to the modulation process.

POWER LEVEL	MAXIMUM LEVEL MEASURED				
	@ 100 kHz	@ 200 kHz	@ 250 kHz	@ 400 kHz	600 kHz to 1.8 MHz
37 dBm	+0.5 dBm	-30 dBm	-33 dBm	-60 dBm	-64 dBm
35 dBm	+0.5 dBm	-30 dBm	-33 dBm	-60 dBm	-62 dBm
$\leqslant$ 33 dBm	+0.5 dBm	-30 dBm	-33 dBm	-60 dBm	-60 dBm

Table 5.3. Specifications for the allowed RF output spectrum due to switching transients process.

POWER LEVEL	MAXIMUM LEVEL MEASURED			
	@ 400 kHz	@ 600 kHz	@ 1200 kHz	@ 1800 kHz
$\leqslant$ 37 dBm	-23 dBm	-26 dBm	-32 dBm	-36 dBm

- the modulation process;
- up and down ramping of the power at the beginning and end of a transmitted frame.

The allowed output RF spectrum due to the modulation process is specified in table 5.2 [ETSI(4.2)]. It gives the maximum allowed level relative to the carrier in function of the frequency from the carrier and the carrier signal level for a 30 kHz bandwidth.

Table 5.3 gives some specifications for the allowed RF output spectrum due to switching transients (up and down ramping) [ETSI(4.2)]. Power levels are again given for a 30 kHz bandwidth.

5.3.3.3 Spurious Emissions. The output RF spectrum specifications of the previous section give some information about transmitted signals which are not situated in

Table 5.4. Measurement bandwidths for the spurious emissions specifications.

FREQUENCY OFFSET	MEASUREMENT BANDWIDTH
IN-BAND FREQUENCIES (890 MHZ - 915 MHZ)	
$\geqslant$ 600 kHz	10 kHz
$\geqslant$ 1.8 MHz	30 kHz
$\geqslant$ 6 MHz	100 kHz
OUT-OF-BAND FREQUENCIES	
$\geqslant$ 2 MHz	30 kHz
$\geqslant$ 5 MHz	100 kHz
$\geqslant$ 10 MHz	300 kHz
$\geqslant$ 20 MHz	1 MHz
$\geqslant$ 30 MHz	3 MHz

the wanted signal band but which are inherently part of the wanted signal due to the effects of either modulation or transient switching. There are however also in-band and out-of-band specifications for any type of spurious signal [ETSI(4.3)].

The power measured in a bandwidth as specified in table 5.4 must for a mobile station, when allocated a channel, be no more than :

- 250 nW in the frequency band from 9 kHz to 1 GHz;

- 1 μW in the frequency band from 1 GHz to 12.75 GHz.

The power measured in a 100 kHz bandwidth must for a mobile station, when not allocated to a channel, be no more than :

- 2 nW in the frequency band from 9 kHz to 1 GHz;

- 20 nW in the frequency band from 1 GHz to 12.75 GHz.

The power emitted by the mobile station must, under all conditions, be lower than 4 pW in a 30 kHz bandwidth in the band from 930 MHz to 960 MHz except for five channels of 200 kHz. In these five channels the power level may be up to 36 dBm.

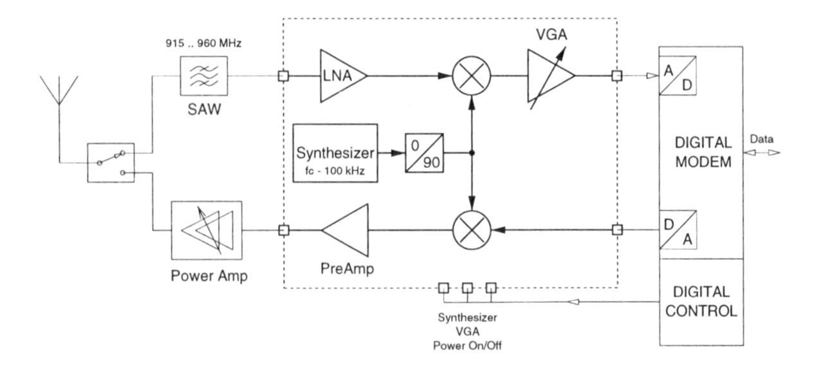

Figure 5.5. The proposed transceiver architecture for GSM.

5.3.3.4 Spurious Emissions of the Receiver. Although not intended to perform any transmission, a receiver can also transmit spurious signals. The spurious emissions of a mobile station receiver, measured in a 30 kHz band, are specified to be lower than [ETSI(5.4)] :

- 2 nW in the frequency band from 9 kHz to 1 GHz;
- 20 nW in the frequency band from 1 GHz to 12.75 GHz.

5.4 A TRANSCEIVER ARCHITECTURE FOR GSM

5.4.1 Block Diagram

Fig. 5.5 gives a schematic representation of the proposed transceiver architecture. It does not show differential and quadrature signals. The proposed architecture is a direct upconversion transmitter, combined with a low-IF receiver. The mirror signal suppression of the receiver is performed in the digital modulator/demodulator part. A highly integrated transceiver chip is proposed : the transmitter and receiver are realized on the same chip in combination with a synthesizer. There is a single synthesizer serving both the receiver and transmitter. This is possible because reception and transmission of frames does not occur at the same time (see fig. 5.2).

With the proposed transceiver chip it is possible to build a complete wireless communication system with only five extra components : an antenna, an antenna multiplexer, a high frequency blocking filter, a power amplifier and a low frequency digital signal processing chip for the demodulation and modulation of the baseband signals.

Fig. 5.6 shows a more elaborate representation of the proposed transceiver chip. It gives an overview of the configuration of all differential and quadrature signals and building blocks. All high frequency signals, except for the local oscillator, are single-ended unbalanced signals. All low frequency and baseband signals are fully

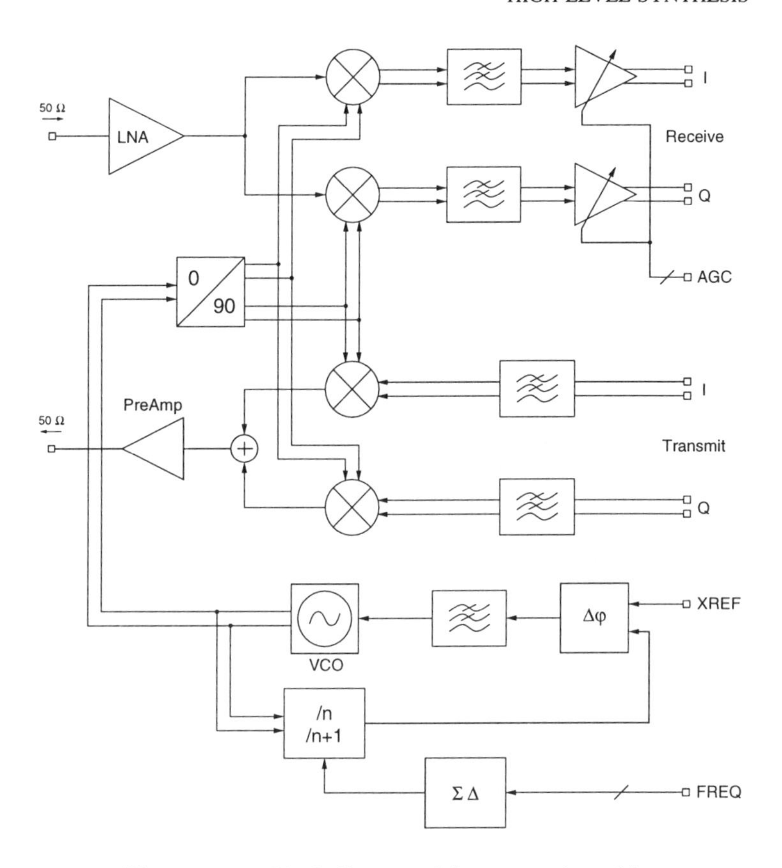

Figure 5.6. Block diagram of the transceiver chip.

differential. The chip has a single RF input for the receiver that has to be driven from a 50 Ω signal source, and a single RF output for the transmitter which can drive a 50 Ω load. There are no external components present in the receive and transmit path, which reduces drastically the complexity and cost of the transceiver. The signal levels for the low frequency input and output are assumed to be 6 dBm, the required high frequency output signal level is assumed to be 0 dBm. For the combined duplexer and high frequency filter a 3 dB signal loss is assumed.

5.4.2 The Receiver

5.4.2.1 The Low-IF Architecture. The proposed receiver uses a low-IF topology. This topology allows for a direct downconversion in a single stage without the disadvantages of the highly sensitive baseband operation, typical for zero-IF direct downconversion receivers.

The single stage downconversion has big advantages. It requires only one mixing stage followed by lowpass filters, resulting in a low power consumption. Single stage downconversion implies that there is no need for filtering in between stages. This reduces the number of external components (no need for ceramic IF filters), the number of external nodes (and thus bondpads) and it reduces the power consumption because external nodes have to be driven at much lower impedances than internal nodes.

The classically used single stage receiver is the zero-IF receiver. The zero-IF receiver however has the big disadvantage that it uses the baseband for its, still weak, wanted signal directly after downconversion. Moreover, parasitic DC signals caused by mismatch, local oscillator self-mixing and RF signal crosstalk also appear in the baseband. These signals vary dynamically under different reception circumstances. It is therefore very difficult to suppress them and separate them from the wanted signals without severe loss of reception quality.

The low-IF receiver uses an, although low, IF frequency. The low-IF receiver uses, like the zero-IF receiver, a single stage quadrature downconversion. The mirror signal suppression is performed at low frequencies, after downconversion, which eliminates the need for a high frequency mirror signal suppression. In the proposed architecture the mirror signal suppression is performed in the digital modem part. A more elaborate discussion of this low-IF architecture was given in section 3.5.1.2.

A careful choice of the low IF frequency allows for a significant reduction of the required building block specifications. The IF frequency is chosen equal to 100 kHz. Fig. 5.7 illustrates that the adjacent channel (at -100 kHz) is then the mirror signal. This signal is defined by the adjacent channel interference specification for the signal on 200 kHz from the carrier. The reference interference performance must be met when the adjacent channel is 9 dB higher than the signal in the wanted channel. The required performance will be met when this mirror signal is suppressed to -9 dB under the wanted signal level. This would thus require an 18 dB mirror signal suppression. However, the mirror signal suppression must be taken higher in order to sufficiently suppress the tail of the adjacent signal at $f_{IF} - 400$ kHz. This adjacent signal can be 41 dB higher than the wanted signal level, but it is not folded on to the wanted signal. Only its tail can interfere with the wanted signal. This interference is limited to the interference of the adjacent signal at $f_{IF} - 200$ kHz when it is suppressed to the same level. The required mirror signal suppression is then :

$$\begin{aligned} &\textit{Required mirror signal suppression} = \\ &\quad\quad 41\text{ dB} \quad \text{[signal at 400 kHz to wanted signal ratio]} \\ &- \quad 9\text{ dB} \quad \text{[signal at 200 kHz to wanted signal ratio]} \\ &= \quad 32\text{ dB} \end{aligned} \tag{5.1}$$

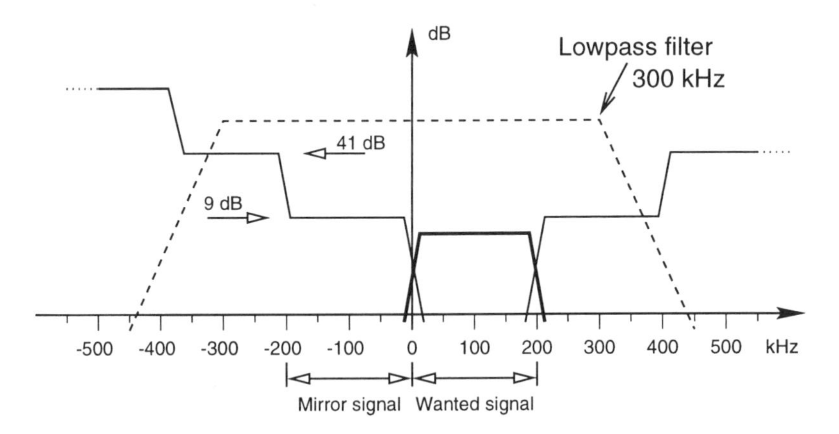

Figure 5.7. The low-IF configuration in the frequency domain after downconversion, showing the wanted signal and its adjacent channels.

This is equivalent with a 1.5° phase accuracy and a 225 mdB relative amplitude accuracy. These are however very safe values which may not be necessary. The interference of the tail is probably very low and a much lower mirror signal suppression may be sufficient. The exact reduction of the interference performance caused by the tail of the signals at 200 kHz and 400 kHz from the carrier can be checked with a dedicated system level design software tool for RF which allows for bit-error-rate simulations. This has however not been done in the work that is presented here. With such software it is possible to determine more exactly the required mirror signal suppression to achieve the specified frame error, bit error and residual bit error rates. The result however will, as shown in the previous paragraph, be lower than 32 dB. However, the quadrature generator is not only used to suppress the mirror signal in the receive path; it is also used to obtain the required single sideband upconversion in the direct upconversion transmitter. The specification for the suppression of unwanted sideband in the transmitter is higher (35 dB) and it is therefore pointless to determine the required mirror signal suppression more accurately by means of BER simulations. A phase accuracy of 1° and 150 mdB is implemented. This is equivalent to a mirror signal suppression (receiver) and an unwanted sideband suppression (transmitter) of 35 dB.

The choice of 100 kHz for the IF also has the advantage that the unwanted transmission (during reception) of the LO signal is situated between two channels. This means that it does not directly degrade the quality of any signal in these channels. Nonetheless, the LO feedthrough to the antenna must be limited to -30 dB.

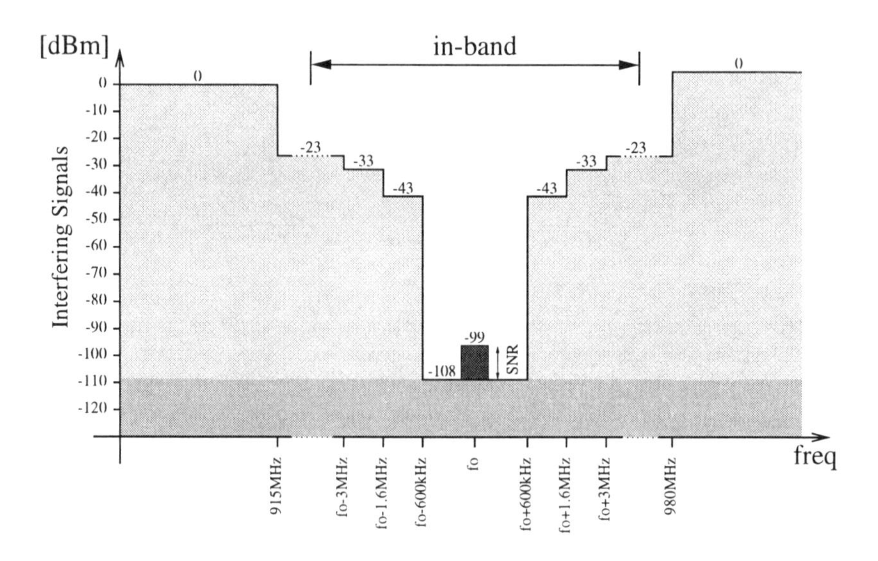

Figure 5.8. The blocking signal levels without HF bandpass and LF lowpass filtering.

5.4.2.2 The Lowpass Filters. The lowpass filtering will be implemented within or directly after the downconverter. This is possible when the filters are limited to a first-order lowpass filter. A corner frequency of 300 kHz $\pm$ 75 kHz, i.e. untuned, is proposed. These lowpass filters introduce, in combination with the automatic gain control (AGC), a further reduction of the dynamic range of the low frequency output signals. The required number of bits for the A/D-converters is directly related to this dynamic range. The dynamic range reduction that can be achieved depends on the order of the lowpass filter that is implemented. Here a very compact implementation with only limited filtering capabilities is chosen, resulting in higher A/D-converter specifications as they will also have to sample the signal in the close neighborhood of the wanted signal. There are two types of signals specified which can be present in the neighborhood of the wanted signal : adjacent channels and blocking signals. The adjacent channels are shown in fig. 5.7, the blocking signals in fig. 5.8.

Fig. 5.7 shows that there can be a large adjacent signal at 400 kHz from the wanted signal channel. This signal is close to the wanted signal frequency and it can therefore not be suppressed with the lowpass filter when a corner frequency of 300 kHz $\pm$ 75 kHz is used. The required signal-to-noise ratio (SNR) for the wanted signal is 9 dB (see section 5.3.2.1), the signal at 400 kHz can, unfiltered, be 41 dB higher than the wanted signal, all signals further away will, due to the lowpass filtering, be less than or almost equal to 41 dB higher (e.g. first-order filtering of a -49 dBm signal at 600 kHz with a corner frequency of 300 kHz gives a 7 dB suppression resulting again in a maximum signal of 42 dB higher than the wanted signal). The overall required dynamic range for the proposed receiver architecture is thus 51 dB (or 9 bits).

At 600 kHz from the wanted signal carrier a blocking signal of -43 dBm can appear. The wanted signal of 99 dBm (3 dB above the reference sensitivity level) must under these conditions still achieve a signal-to-noise ratio of 9 dB. The required dynamic range would be 65 dB (i.e. 11 bits). A first-order lowpass filter with a worst case corner frequency of 375 kHz will attenuate the blocking signal at 600 kHz with 4.5 dB to -47.5 dBm and all blocking signals further away will, due to the lowpass filtering, be less than -47.5 dBm. The required dynamic range is thus 60.5 dB (i.e. 11 bits).

With higher order filtering the dynamic range can be reduced down to a theoretical value of 30 dB (or 5 bits) (this is the dynamic range of the wanted signal only for all possible signal levels). However, the cost of an 11 bit A/D is not high anymore, especially when a delta-sigma converter is used (because a conversion bandwidth of 200 kHz is then sufficient) and it is therefore better to implement a transceiver with a lower complexity (an untuned first-order lowpass filter) at the cost of a better A/D-converter (11 bit at 200 kHz).

The use of a lowpass filter with a nominal corner frequency of 300 kHz allows for an untuned implementation. An untuned implementation will result in a corner frequency variation (assuming 25 % variation in both directions due to capacitor and resistor variations) from 225 to 375 kHz. As shown above, 375 kHz will still give sufficient suppression of the adjacent signals. A 300 kHz corner frequency is situated 100 kHz from the closest possible wanted signal component (situated from near DC to 200 kHz), resulting in this way in a minimal in band phase and amplitude distortion for the wanted signal.

5.4.2.3 The AGC. The dynamic range of the low frequency output signals is not only determined by the ratio between the smallest possible wanted signal and the largest possible adjacent signals after filtering. The ratio between the smallest and the largest possible wanted signal is also important. This ratio gives a dynamic range of :

Required dynamic range without AGC =

	-12 dBm	[largest possible wanted signal]
−	(-102 dBm)	[smallest possible wanted signal]
−	9 dB	[required SNR on the smallest possible wanted signal]
=	99 dB	

(5.2)

This is a very large value. Because both cases can not appear at the same time, this dynamic range can be greatly reduced (down to the 60.5 dB spec of the previous paragraph) by using automatic gain control.

A high signal in the adjacent channels will limit the maximum applicable gain to the wanted signal to :

$$\begin{array}{lll} \multicolumn{3}{c}{\textit{Required gain when adjacent channels are present} =} \\ & 6 \text{ dBm} & \text{[output signal level]} \\ - & (-43 \text{ dBm}) & \text{[largest possible adjacent signal]} \\ - & (-3 \text{ dB}) & \text{[signal loss in the duplexer]} \\ = & 52 \text{ dB} & \end{array} \tag{5.3}$$

For the smallest possible wanted signal the maximum applicable gain is only limited by the blocking levels :

$$\begin{array}{lll} \multicolumn{3}{c}{\textit{Required gain for the smallest possible wanted signal} =} \\ & 6 \text{ dBm} & \text{[output signal level]} \\ - & (-47 \text{ dBm}) & \text{[largest possible blocking signal]} \\ - & (-3 \text{ dB}) & \text{[signal loss in the duplexer]} \\ = & 56 \text{ dB} & \end{array} \tag{5.4}$$

18 dB of this gain is realized in the LNA, 38 dB must be realized during and after downconversion.

The largest possible wanted signal is -15 dBm (i.e. -12 dBm - 3 dB signal loss in the duplexer). It requires an overall gain of :

$$\begin{array}{lll} \multicolumn{3}{c}{\textit{Required gain for the largest possible wanted signal} =} \\ & 6 \text{ dBm} & \text{[output signal level]} \\ - & (-15 \text{ dBm}) & \text{[largest possible wanted signal signal]} \\ = & 21 \text{ dB} & \end{array} \tag{5.5}$$

The 18 dB of the LNA must under these conditions be switched off (to prevent blocking at the input of the downconverters) and the gain after downconversion must thus be 21 dB.

During and after downconversion a variable and controlable gain must be realized which can vary between 21 and 38 dB. This gain can be controlled by the digital modem chip by measuring the signal level after A/D-conversion. The gain of the LNA must be switchable between 0 and 18 dB.

5.4.2.4 DC Blocking Filters. Parasitic DC signals, mainly due to crosstalk between the mixer inputs, will still be produced in the low-IF receiver topology. The difference with the zero-IF receiver is that they do not interfere anymore with the wanted signal and that they can therefore be suppressed without degradation of the wanted signal. The proposed low-IF approach uses only a limited low frequency filtering (first order), also resulting in a limited gain (max. 56 dB) in the receive path. Different from the zero-IF receiver (where up to 5th order lowpass filtering and 80 dB gain is used) this means that the parasitic DC signals will not saturate the lowpass filters and that they can be sampled. The maximum DC signal that can appear after downconversion, lowpass filtering and amplification with the AGC is about -6 dBm. The following A/D-converter can sample +6 dBm, so the DC-signal can be sampled by increasing the dynamic range of the A/D-converters with 2 dB (i.e. 1/3 of a bit). The DC-signals can then be suppressed in the digital part. By postponing this suppression until after the final downconversion to the baseband (performed in the DSP, see section 5.4.5), the DC signal suppression will be done intrinsically with the final 100 kHz lowpass filter (the DC signal is upconverted to 100 kHz in the final downconverter).

5.4.2.5 The HF Bandpass Filter before the LNA. A high frequency, high Q passive filter must be placed before the LNA. Without filtering, there can be no gain implemented in the LNA. An out-of-band signal of 0 dBm can be present and any gain would therefore saturate the input of the downconverters. When no filtering is used, direct downconversion (i.e. without first amplifying the antenna signal) is necessary. This would require from the downconversion mixers an input dynamic range of :

Required input dynamic range =

	0 dBm	[highest possible blocking level]
−	(-102 dBm)	[smallest possible wanted signal level]
−	9 dB	[required SNR on the smallest possible wanted signal level]
=	111 dB	

(5.6)

And the noise figure (NF) for the downconverters must be equal to the NF usually required from the LNA. These are specifications which are very hard to achieve, if not impossible.

When a SAW (standing acoustic wave) filter with a bandwidth equal to the GSM mobile station receive band is used, the out-of-band blocking signals are suppressed and an LNA can be used. The maximally allowable gain for the LNA is equal to :

Maximum allowable LNA gain =

	0 dBm	[highest allowed downconverter input signal]
−	(-23 dBm)	[maximum in-band blocking signal]
−	(-3 dB)	[signal loss in the duplexer]
=	26 dB	

(5.7)

A high gain is however hard to achieve at high frequencies and a lower LNA gain may be used at the cost of a lower NF spec for the downconverters. The use of a blocking filter will introduce some signal loss. A maximum of 3 dB signal loss for the duplexer - blocking-filter combination can be accepted.

A second reason for using a SAW filter is the parasitic mixing with higher order harmonics of the oscillator. Without filtering, the antenna signal is broadband and a 0 dBm signal at e.g. 2.7 GHz can be mixed with e.g. a 30 dB suppressed second-order LO harmonic, resulting in a -30 dBm signal. This signal would, after mixing, be superimposed on a -99 dBm wanted signal. A strong suppression before mixing of the signals at 2.7 GHz is thus very important.

5.4.2.6 The LNA. The gain in the low noise amplifier (LNA) may, according to equation 5.7, be maximally equal to 26 dB. High gains are however difficult to achieve at high frequencies. The gain in the LNA is therefore set to 18 dB. Under these circumstances the largest possible blocking signal after the LNA is equal to -8 dBm. The largest possible wanted signal (-15 dBm) would be amplified to +3 dBm which is a too large value for both the output of the LNA and the input of the multipliers. The gain of the LNA must therefore be switchable between 0 dB (used under the condition of a large wanted signal) and 18 dB (used under the condition of a low input signal).

The required total equivalent input noise figure (NF) which must be achieved at all times at the antenna input (i.e. before the duplexer) is equal to :

Required total equivalent input NF =

	-102 dBm	[smallest possible wanted signal]
−	9 dB	[required SNR on smallest possible wanted signal]
−	(-173.9 dBm + 53.0 dB)	[the noise of 50 Ω in 200 kHz]
=	9.9 dB	

(5.8)

An extra margin must be taken on this value in order to meet the required specification even when accumulation with other noise sources (like the downconversion mixers and the VCO) occurs [Bird RFD93]. The total equivalent input noise figure spec is therefore set to 9 dB (maximum value). The loss in the duplexer and blocking filter is assumed to be typically 3 dB, maximum 3.5 dB. The noise figure for the LNA is set to 4 dB (maximum value).

5.4.2.7 A Bandpass Filter between LNA and Downconverter. The use of a bandpass filter between the LNA and the downconverter is not necessary when a highly linear LNA and a non-switching downconversion mixer are used. A bandpass filter after the LNA suppresses the second and third order harmonic components of the RF signal (at 1.8 and 2.7 GHz) which are generated within the LNA due to a limited linearity. A switching mixer (multiplication with a square or near square wave) would bring these unwanted harmonics down to the same IF frequency as the wanted signal. The use of a CMOS technology allows for the realization of a highly linear LNA (2nd and 3rd harmonic components < -50 dB) and a highly linear non-switching mixer (2nd and 3rd LO harmonic < -30 dB), resulting in a sufficient suppression of the unwanted mixing products.

From the above specifications the required input third order intercept point performance, IP3, for the LNA and the mixer can be calculated. 2nd harmonic distortion can be sufficiently suppressed by using a fully balanced LO signal. The required third order distortion spec, HD3, will be reduced because of the filtering effects which will occur at 2.7 GHz. Assuming a value of 10 dB filtering, the HD3 specification becomes 40 dB. The third order intermodulation, IM3, is 10 dB higher than HD3, resulting in an IM3 of 50 dB for an input signal of -26 dBm. This is equivalent to an IP3 of -1 dBm ($= -26$ dBm $+$ 50 dB/2). The mixer must then have an input IP3 of 17 dBm (i.e. -1 dBm + 18 dB LNA gain).

The elimination of the filter between the LNA and the downconverter offers many benefits. It reduces the number of external components, and thus the cost, and it also reduces the power consumption, because going off-chip to drive the external filter requires a matching circuit which can drive a 50 Ω load. The extra advantage is that the linearity of the LNA can be much higher when the impedance which it has to drive is higher (e.g. 250 Ω).

5.4.2.8 The Downconverter. The NF required from the components after the LNA added to the noise of the LNA (5 dB), must be lower than the required equivalent input noise figure before the LNA under worst case conditions (i.e. 5.5 dB, or 9 dB - 3.5 dB, the duplexer loss). This gives an equivalent NF before the LNA for the components after the LNA of 3.1 dB. The required NF of the downconverter, lowpass filters and variable gain amplifiers, is therefore :

$$\begin{aligned} &\textit{Required NF for components after the LNA} = \\ &\quad 3.1 \text{ dB} \quad [\text{excess NF}] \\ + &\quad 18 \text{ dB} \quad [\text{LNA gain}] \\ = &\quad 21.1 \text{ dB} \end{aligned} \tag{5.9}$$

A safety margin should be taken on this noise figure. The noise figure of the downconverter is therefore set to 20 dB (maximum value). This value results in a total equivalent input noise for the complete receiver (i.e. LNA and downconverter) at the antenna input (before the duplexer) of 6.6 dB NF (the required value is 9.9 dB, see equation 5.8). The noise figure of a quadrature downconverter is 3 dB higher than the noise figure of its individual downconverters (see section 4.2.2.2). The required NF for each individual downconversion path is therefore 17 dB.

5.4.3 The Synthesizer

5.4.3.1 The VCO. A low phase noise voltage controlled oscillator (VCO) is an essential building block for the transceiver. The phase noise specification is determined by the unwanted downconversion of adjacent channels. For the adjacent channel at 400 kHz this gives :

Required phase noise due to adjacent channels =

	-41 dB	[adjacent signal at 400 kHz to wanted signal ratio]
−	9 dB	[required SNR on smallest possible wanted signal]
−	53.0 dB	[200 kHz bandwidth]
=	-103 dBc/Hz @ 400 kHz	

(5.10)

For the blocking signal at 600 kHz this gives :

Required phase noise due to blocking signals =

	-102 dBm	[smallest possible wanted signal]
−	9 dB	[required SNR on smallest possible wanted signal]
−	(-43 dB)	[blocking signal at 600 kHz]
−	53.0 dB	[200 kHz bandwidth]
=	-121 dBc/Hz @ 600 kHz	

(5.11)

The latter is more stringent and therefore taken as specification [Cran CASII95]. An extra margin of e.g. 4 dB must be taken on this value.

5.4.3.2 The PLL. Two aspects determine the design of the PLL : the phase noise of the synthesized signal and the transient synthesizer requirements such as settling time [Cran CASII95, Gard 1979]. Noise arises from three sources : the phase noise of the reference source, the phase noise of the VCO and the phase noise introduced by the fractional-N synthesis with sigma-delta (Σ/Δ) techniques. The Σ/Δ technique is proposed because it gives a much better performance than the classically used accumulators [Riley JSSC93, Riley CASII94].

The prescaler must have a tuning range equivalent to the required VCO tuning range. Since the phase noise of the reference is multiplied with the division factor, this number must not be too high. However, the reference frequency must not be too high either, in order to allow an easy implementation of the Σ/Δ modulator and the phase detector. The PLL requires the following blocks : a divide by 64+N counter, a 26.0 MHz Xtal clock reference, a phase detector, a loop filter and a sigma-delta modulator.

5.4.3.3 The Quadrature Generator. The quadrature generator must produce quadrature output signals with a phase accuracy of less than 1° and a relative amplitude accuracy of less than 150 mdB (requirements set by the low-IF downconversion topology and the direct upconversion transmitter topology, see section 5.4.2.1) in a broad passband around 900 MHz. The broad passband is required to serve both the reception and the transmission path and to avoid any necessity for tuning or trimming. A large bandwidth can be achieved by using a second-order quadrature generator structure.

5.4.4 The Transmitter

5.4.4.1 The Upconverter. The proposed transmitter uses a direct upconversion architecture. In this topology baseband signals are directly upconverted to the wanted carrier frequency by means of quadrature multiplication. This ensures very low out-of-band spurious signals and it requires no external high frequency filter for mirror signal suppression. In this topology the carrier frequency is equal to the local oscillator frequency and special attention must be given to the possible influence of the antenna signal with its high output power level on the PLL control phase locked loop loop. This effect is minimized by using a separate power amplifier which is implemented on a different die and by using at the same time a fully integrated VCO. In order to have a very good isolation not only the vari-cap is integrated on the die with the VCO, but the complete LC-tank (the inductor included) is integrated with the VCO [Cran VLSI96]. In that case the VCO has not a single external pin with a high frequency signal on, making it very insensitive to the signal at the output of the power amplifier on a different die. Nonetheless special care must be given to the design and layout of a VCO with very low sensitivity, e.g. by using large signals (+6 dBm) in the VCO and at the input of the prescaler/counter. Special care will also have to be given to a very good decoupling of power supply and ground nodes.

Both the phase and amplitude information of the wanted signal, together with the frame shape information (for up and down ramping) is generated in the digital modem chip and synthesized with a 6 bit D/A-converter. Control of the output power level and suppression of second and third harmonic distortion components is performed in the power amplifier. The power amplifier is driven from a preamplifier which delivers a 0 dBm output signal to a 50 Ω load at the input of the power amplifier.

The required SNR for the transmitted signals is 30 dB. This implies a quadrature accuracy of less than 1.5°. An extra margin of 5 dB is taken for this value. This sets the quadrature accuracy requirements to a 1° phase error and an amplitude error of 150 mdB. Third other intermodulation must be smaller than -30 dB and the input dynamic range must be higher than 30 dB.

5.4.4.2 The Anti-Aliasing Filter. The baseband quadrature signals, synthesized with the D/A-converters in the digital modem chip, have to be lowpass filtered for anti-aliasing before upconversion. Extra filtering is necessary when the input bandwidth of the upconversion mixers is not low enough to perform this task. Especially when delta-sigma D/A-conversion is used, this anti-aliasing filtering can be of a first-order untuned passive nature which can be easily implemented in the digital modem chip.

5.4.4.3 The Preamp. The preamp must convert a 0 dBm signal from a high impedance (e.g. 250 Ω) to a 50 Ω output impedance. Its most important specifications are input bandwidth (200 kHz) and IM3 (< -30 dB).

5.4.5 The Digital MODEM Chip

The digital modemchip has to perform the following functions :

- *peak value measurement of the sampled receive signal and control of the analog VGA;*
- *final downconversion of the receive signal at low IF to the baseband;*
- A/D and D/A with anti-aliasing;
- high order lowpass filtering of the baseband receive signal;
- demodulation of the baseband quadrature receive signal;
- baseband synthesis of the quadrature signal to be transmitted.

Only the functions in italics are different from the classical digital modem chips for zero-IF or zero-IF like receivers. The digital circuitry must of course also control the signal flow (frequency and time allocation), output levels and switching of the antenna signal, but this has no influence on the design of the analog transceiver front-end part as proposed here.

5.4.5.1 The Final Downconversion. The block diagram for the digital downconversion from the 100 kHz IF frequency to the baseband, is given in fig. 5.9. The baseband signal, generated in the analog front-end is sampled with a 10 bit accuracy. The wanted information is still centered around the 100 kHz IF frequency, together with the mirror signal which is not yet suppressed and which has to be suppressed in the digital modem chip. The input signal is still a broadband signal (up to several MHz), but its dynamic range is guaranteed to be, for all frequencies, lower than 66 dB (see section 5.4.2.2). Only the lowest baseband signal, from DC to 200 kHz, has to be sampled with this accuracy. For all other signals it is only important that they do not saturate the A/D-converter and that they do not give aliasing components into the wanted signal band. Saturation can not occur as all signals have a dynamic range lower than 66 dB. Aliasing can occur when non sufficient anti-aliasing filtering is performed.

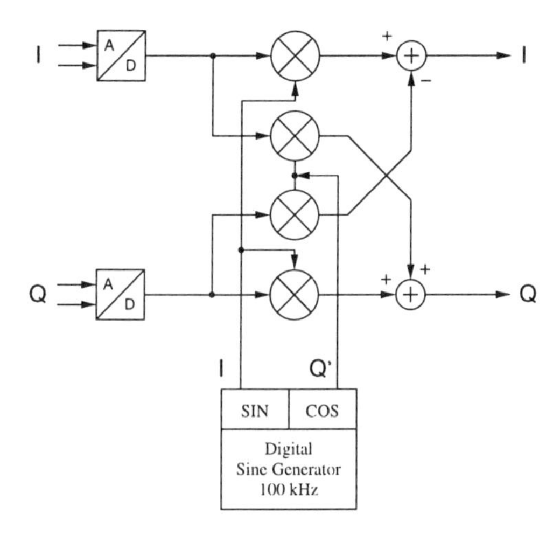

Figure 5.9. Final downconversion to baseband in the digital modem chip.

How much anti-aliasing has to be performed depends on the used oversampling ratio in the A/D-converters. A Σ/Δ-converter would have a very high oversampling ratio and a first-order anti-aliasing filter would suffice. This requires however the use of a decimation filter which has to be implemented in the digital signal processing part [Optey KUL90]. A trade-off can therefore be made in the digital design and the choice of the A/D-converter.

The final downconversion is performed in the digital modem chip by means of a double quadrature multiplication of the sampled signal with a quadrature sinusoidal signal of 100 kHz. This will downconvert the wanted signal to the baseband, centered around DC, and it will upconvert the mirror signal to a center frequency of 200 kHz. This means that after this operation they can be separated by means of lowpass filtering. For the actual implementation of the sine generator and the multipliers many trade-offs can be made. A lot depends on the used clock frequency and the ratio between the clock frequency and the IF of 100 kHz. The use of a $\Sigma\Delta$-converter for the A/D-converter results, before decimation, in a bit stream signal. This has the advantage that multiplication of this bit stream signal with a sine and cosine reduces to the inverting and non-inverting of numbers. Decimation can then be performed after the final downconversion together with the lowpass filtering. Again, many trade-offs can be made in the design of the digital signal processor.

5.4.5.2 The Digital Lowpass Filter. After the final downconversion all unwanted signals (mirror signal, adjacent channels and the upconverted parasitic baseband signals) must be fully suppressed. This can be done with two high order (e.g. 64 or 128

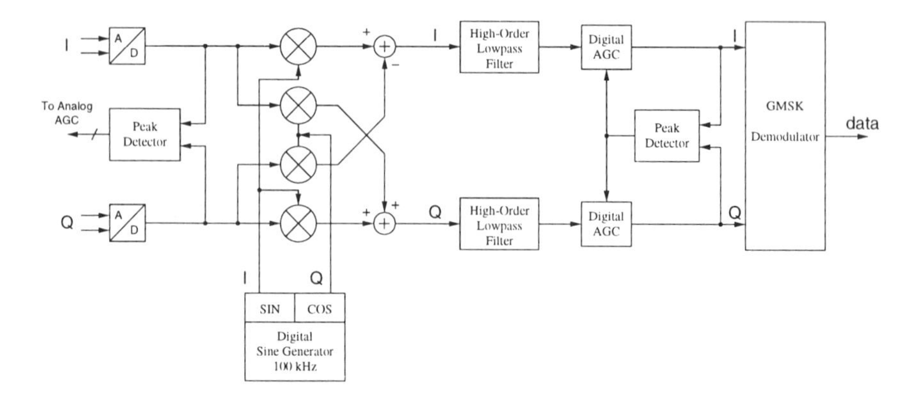

Figure 5.10. Peak detection, gain control and lowpass filtering in the digital modem chip.

taps) digital lowpass filters with a corner frequency of 100 kHz. Fig. 5.10 shows the position of these lowpass filters. These filters are not different from the lowpass filters used in any other receiver topology. The corner frequency has to be taken slightly lower (e.g. 98 kHz) in order to also fully suppress the upconverted DC-signals which are now situated at 100 kHz.

5.4.5.3 Peak Detection and AGC Control. After each filter operation the signal levels must be adjusted to the input signal capability of the succeeding stages. This is done by means of two AGC functions : an analog-digital AGC for the first lowpass filtering and a fully digital AGC for the final high order lowpass filtering. The analog-digital AGC is an analog VGA amplifier controlled from a digital peak-detector. The VGA can be changed in steps of 6 dB. The gain must be increased with 6 dB when the maximum of the peak values of the sampled I and Q signal is lower than -8 dB below the input signal capability of the A/D-converters (+6 dBm). The gain must be decreased with 6 dB when the maximum of the peak values of the sampled I and Q signal is higher than 1 dB below the input signal capability of the A/D-converters. The gain of the LNA (18 dB) must be switched off when the gain of the analog VGA can not be decreased any further while, according to the previous rule, it should. The gain of the digital AGC must be able to be varied between 0 and 55 dB.

5.4.6 Overview of Circuit Level Specifications

Fig. 5.11 shows the obtained level diagram for the proposed low-IF receiver architecture. The signals in fig. 5.11 are :

(1) : The highest possible wanted signal

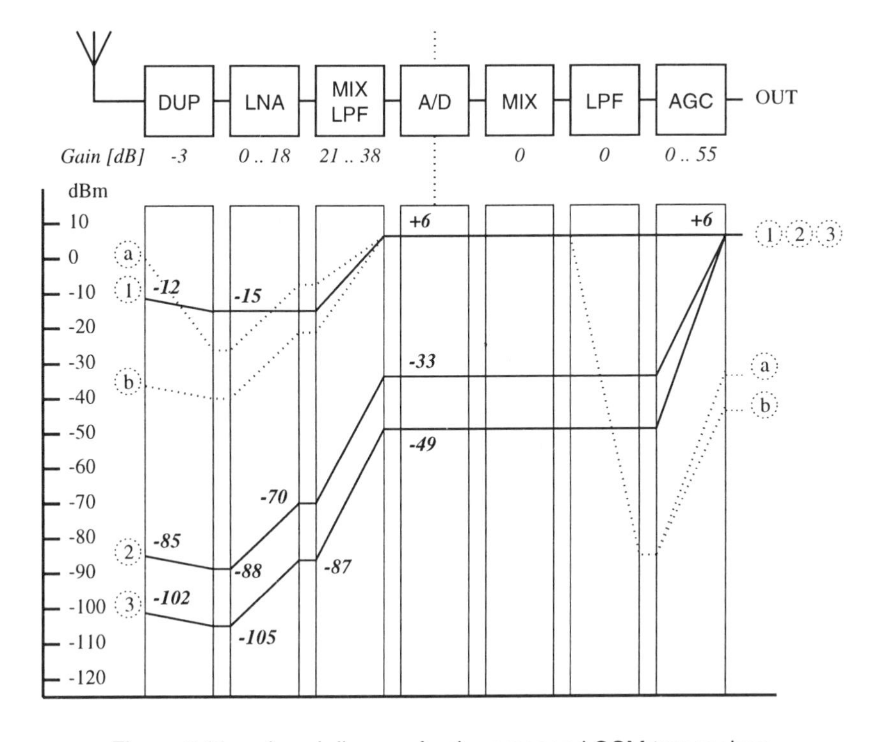

Figure 5.11. Level diagram for the proposed GSM transceiver.

(2) : The smallest possible wanted signal when high adjacent channels are present

(3) : The smallest possible wanted signal

(a) : The highest possible blocking level under condition (3)

(b) : The highest possible unwanted adjacent signal under condition (2)

Tables 5.5, 5.6 and 5.7 give an overview of the derived building block specifications for respectively the receiver, the synthesizer and the transmitter of the proposed transceiver chip for GSM.

5.5 CONCLUSION

In this chapter the full high-level design of a GSM transceiver front-end was given as a practical illustration of the new design techniques which where proposed in chapters 3 and 4. As receiver architecture a low-IF receiver with the final downconversion performed in the baseband, was proposed. On the one hand, this architecture gives a good

Table 5.5. Overview of the derived building block specifications for the receiver of the proposed GSM transceiver chip.

Receiver		
BUILDING BLOCK	SPECIFICATION	VALUE
HF bandpass filter	Passband	935 MHz to 960 MHz
	Signal loss	$\leqslant$ 3.0 dB, typical
LNA	Gain	0 dB or 18 dB, switchable
	Bandwidth	> 1 GHz
	Output signal capability	-8 dBm
	Input signal capability	-26 dBm
		-13 dBm with 0 dB gain
	Input IP3	-1 dBm
	Noise figure	< 4 dB with 18 dB gain
Downconversion mixers	IF frequency	100 kHz
	Input bandwidth	1 GHz
	Output bandwidth	see lowpass filter
	Input signal capability	-8 dBm
	Input IP3	13 dBm
	Output signal capability	see VGA
	LO mixing signal	0 dBm
	Conversion gain	21 dB, part of the VGA gain
	Noise figure	< 17 dB
Lowpass filters	Corner frequency	300 kHz $\pm$ 75 kHz
Variable gain amplifiers	Gain	21 to 38 dB, in steps of 6 dB
	Bandwidth	see lowpass filter
	Output signal capability	6 dBm
A/D-converters	Input bandwidth	200 kHz
	Dynamic range	11 bit

Table 5.6. Overview of the derived building block specifications for the synthesizer of the proposed GSM transceiver chip.

Synthesizer		
BUILDING BLOCK	SPECIFICATION	VALUE
VCO	Tuning range	890 MHz to 960 MHz
	Phase noise	-121 dBc/Hz @ 600 kHz
Phase locked loop	Reference frequency	26 MHz
	Frequency divider	64+N counter, with N = 1..9
Quadrature generator	Center frequency	900 MHz
	Bandwidth	> 200 MHz
	Output signal	0 dBm
	Phase accuracy	$< 1^{o}$
	Amplitude accuracy	< 150 mdB

Table 5.7. Overview of the derived building block specifications for the transmitter of the proposed GSM transceiver chip.

Transmitter		
BUILDING BLOCK	SPECIFICATION	VALUE
Upconversion mixers	Output bandwidth	1 GHz
	Input bandwidth	> 150 kHz
	LF input signal	+6 dBm
	Output signal	0 dBm
	IM3	< -30 dBm
	Oscillator feedthrough	< -30 dBm
	LO mixing signal	0 dBm
	Conversion gain	-6 dBm
Preamp	Bandwidth	1 GHz
	IM3	< -30 dB
	Input signal capability	0 dBm (high impedance)
	Output signal capability	0 dBm
	Gain	0 dB
	Output driving capability	50 Ω

example of the high degree of integration that can be achieved by using the complex signal techniques to synthesize the receiver and transmitter architecture. On the other hand, this architecture gives also a good insight in how, with an ever further evolution of A/D-converter technology, the interface between analog and digital signal processing can be placed closer to the antenna. The proposed high-level design for a GSM transceiver front-end is therefore a representative example of how transceiver design for future generations wireless applications will be different from the classically used techniques.

6 BUILDING BLOCKS FOR CMOS TRANSCEIVERS

6.1 INTRODUCTION

Today, transceiver implementations usually use a combination of different technologies : CMOS for the baseband digital signal processing, silicon bipolar or BiCMOS for the downconversion, upconversion and IF signal processing and for instance GaAs for the LNA, the duplexer and the power amplifier. The level of integration that is reached in transceiver design is therefore, especially when compared to other integrated telecommunication realizations, still very low. Market demands are however for a lower power consumption, a lower weight and physical volume and a lower cost. These demands require an evolution to a much higher degree of integration. This can be achieved by using new receiver and transmitter topologies (see chapter 3), but it also requires a unification of the technology used for the design of transceiver building blocks.

A transceiver's baseband DSP consists of an enormous amount of digital gates (up to several hundred thousand gates). On itself it occupies more than 75 % of the totally used chip area in a transceiver. It is therefore obvious that it has to be realized in the technology which offers the highest possible density at the lowest possible power consumption and cost price. This means that for the DSP state-of-the-art CMOS technologies have to be used. The use of any other technology, including BiCMOS technologies, would result in a significant reduction of the performance, an increase of the required chip area and a significant increase of the chip cost. Often it is argued that the cost price per area of a BiCMOS technology is only slightly (e.g. 50 to 100 %) higher than the price per area of an advanced sub-micron CMOS process, which would make BiCMOS the preferred technology for RF integration because of the higher degree of integration and the faster design cycle that can be achieved. Today, when looking at the single-chip integration of the analog transceiver front-end, this is true, especially because the chip area required for the implementation of analog circuits does not depend so much on the minimal gate lengths of the MOS devices in the used technology. However, as stated before, BiCMOS can not remain the preferred technology in the future if one is going to consider the single-chip integration of the full transceiver. A BiCMOS technology has only advantages for the analog front-end part which is only 25 % or less of the full transceiver chip area. For the remaining 75 % of digital circuitry it has no advantages. In fact, the area and cost of a DSP increases significantly when it is realized in BiCMOS instead of the most advanced CMOS process. This means that either the idea of full transceiver integration must be abandoned or new design techniques will have to be developed which will allow the implementation of RF circuits in the preferred process for the DSP, the most advanced sub-micron and deep sub-micron CMOS processes.

There is of course one major drawback when using a CMOS technology for the realization of RF circuitry : the poor performance of MOS devices compared to bipolar devices. Bipolar devices can have, for the same current, a 5 to 10 times higher g_m than nMOS transistors. This is a property which will not change with the further development of CMOS technologies towards deep sub-micron gate lengths. This MOS property is mainly a drawback for the realization of output buffers which have to drive signals off-chip into a 50 Ω load. These are power consuming building blocks and a good power efficiency is therefore one of their main requirements. The use of new receiver and transmitter topologies that can be highly integrated, reduces this problem considerably. IF stages are omitted (see chapter 3), and in this way the number of high frequency output nodes terminated to 50 Ω is reduced from 3,4 or 5 to only one, i.e. the output of the transmitter.

A more important problem is the poor frequency performance of a MOS device compared to a bipolar device. A MOS device with a 5 times lower g_m/I ratio than a bipolar transistor, will on top of that also have a lower f_T. A bipolar transistor can have an f_T of 30 GHz. In today's sub-micron technologies, an nMOS transistor has at a low $V_{GS} - V_T$ (i.e. in the region where it has a good power efficiency) an f_T of maximally 10 GHz. The further downscaling of CMOS technologies to deep sub-micron gate lengths, like 0.15 μm, will however resolve this problem. f_T's of 100 GHz have already been reported [Gray CICC95] and because of the much higher investments made for the development of CMOS processes than bipolar, it may be expected that bipolar and BiCMOS technologies will not be able to follow this trend. With f_T's of nMOS devices getting close to the f_T's of npn devices, deep sub-micron CMOS technologies will become a viable alternative for BiCMOS technologies to do system integration. In fact, some silicon foundries may decide not to develop a BiCMOS technology anymore derived from their deep sub-micron CMOS process. Instead they may choose to develop only an analog version (i.e. add only high quality resistors and capacitors) of their highly advanced CMOS process, making it in this way suited for the realization of advanced full system integration.

In this chapter the realization of analog transceiver front-end building blocks in CMOS is studied. The goal is to demonstrate that RF circuits realized in deep sub-micron CMOS technologies can be an alternative for bipolar RF circuits and that they can achieve equal or better performances. For this, new circuit techniques have to be developed which exploit the advantages of CMOS and which overcome its disadvantages. It is not the intention to obtain improved results based on specific features of newly available CMOS technologies. Instead, circuit design techniques are introduced which depend on general properties of MOS devices and chip realizations are made in a wide variety of sub-micron CMOS technologies that have passive components available. In this way it is assured that the circuit design techniques for CMOS RF integration which introduced and demonstrated in this chapter will also hold for future generations of sub-micron and deep sub-micron technologies.

6.2 CMOS MIXERS

6.2.1 Multiplying in CMOS

The linearity of an RF mixer is in most cases rather limited. The Gilbert topology is the most common used in a bipolar technology [Gilb JSSC68, Gilb ISSCC83]. Its operation is based on a translinear configuration, i.e. the use of the exponential voltage to current conversion of the bipolar transistor. Techniques like predistortion and emitter degeneration are necessary to obtain a reasonable linearity. Imperfections in this structure combined with a limited matching will render a third-order intercept point (IP3) which can only be slightly larger than 0 dBm [Meyer JSSC86]. In CMOS a double balanced structure which cancels out the quadratic term of the MOS transistor can be used [Baban JSSC85, Song JSSC90]. These mixers not only have a limited linearity which highly depends on matching, even more important is their limited frequency range. The input transistors of these mixers can only be biased with a relatively small $V_{GS} - V_T$ in order to keep them in saturation at all times. The result is large input transistors which limit the maximal achievable input bandwidth to about 100 MHz.

A better solution is to use the transistors in the linear region, preferably with a large $V_{GS} - V_T$. In this way a small R_{DS} can be realized with a small transistor, allowing a high input bandwidth. This property has been used to realize high frequency GaAs commutating mixers [Rober 1989]. The transistors are being used as pass-transistors and therefore the linearity of these mixers does not depend on the voltage-to-current conversion characteristic of the transistors. The linearity is mainly determined by the speed limitation of the pass-transistors and by the generation of spurious signals during switching. This technique can also be used in CMOS. The output signal after commutation is however still a high frequency signal and it can therefore not be further processed in CMOS. However, receivers only require a downconversion mixer and this means that the output bandwidth may be limited. A solution based on subsampling with a switched-capacitor structure has been proposed in [Chan ESSC93, Shen ISSCC96, Sheng ISSCC96]. The MOS-transistors are used as pass-transistors and an IP3 of 27 dBm has been reported. Drawback of subsampling is the high noise level that is obtained due to aliasing.

A CMOS mixer with very high linearity can be realized by using the linear voltage-to-current characteristic of the MOS transistor in its triode region [Song JSSC86]. The use of a double balanced structure cancels out the common-mode DC biasing signals and the non-linear dependency of g_{DS} on V_{DS}. The remaining problem is still the further processing of the high frequency output current. This limits in [Song JSSC86] the bandwidth to 200 MHz at the cost of a high power consumption and a reduced linearity of the output stage.

6.2.2 A Highly Linear CMOS Downconversion Mixer

Fig. 6.1 presents a new downconversion mixer topology with a very high linearity. The proposed topology is based on the topology of [Song JSSC86] and [Czarn CAS86] which use a cross-coupled structure of 4 nMOS transistors modulated in their linear region. There is however a very important difference. Two extra capacitors have been added on the virtual ground nodes of this topology (the capacitors C_v in fig. 6.1). Indeed, the output stage, the opamp and the feedback resistors, which convert the output current of the mixing transistors back into a voltage, must, in a downconversion mixer, only be able to produce low frequency output signals. However, in order to let the input structure operate correctly for all frequencies, there may not be a signal on the virtual ground nodes of the mixer at any time. This is normally obtained with the feedback structure over the opamp which creates the virtual ground at its inputs, but when the op-amp is only capable of creating this virtual ground for low frequencies the transistors in the input structure will operate as pass-transistor for high frequency signals. It is for this reason that the capacitors C_v have been added. They ensure that all high frequency currents injected to the virtual ground nodes are filtered out and not converted into voltages. At low frequencies, the opamp still generates the virtual ground. By using the structure of fig. 6.1, it becomes possible to split the design of the input structure and the opamp. The input structure can now be optimized for operation at very high frequencies (more than 1 GHz), while the opamp can be designed for low frequency operation (a few MHz).

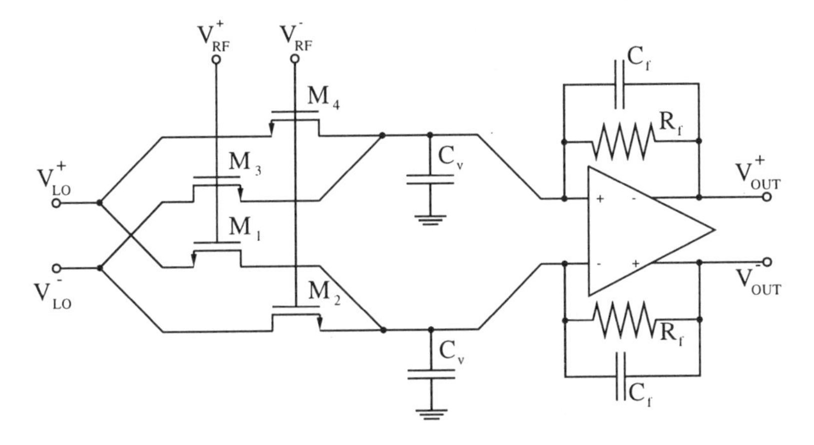

Figure 6.1. The proposed topology for a highly linear downconversion mixer.

With a perfect virtual ground, the currents through the modulated transistors are given as [Laker 1994] :

$$
\begin{aligned}
I_{DS,1} &= \beta_1 \cdot \left(V_{RF}^+ - V_{LO,DC} - V_{Tn,1} - \frac{V_{LO}^+ - V_{LO,DC}}{2} \right) \cdot \left(V_{LO}^+ - V_{LO,DC} \right) \\
I_{DS,2} &= \beta_2 \cdot \left(V_{RF}^- - V_{LO,DC} - V_{Tn,2} - \frac{V_{LO}^- - V_{LO,DC}}{2} \right) \cdot \left(V_{LO}^- - V_{LO,DC} \right) \\
I_{DS,3} &= \beta_3 \cdot \left(V_{RF}^+ - V_{LO,DC} - V_{Tn,3} - \frac{V_{LO}^- - V_{LO,DC}}{2} \right) \cdot \left(V_{LO}^- - V_{LO,DC} \right) \\
I_{DS,4} &= \beta_4 \cdot \left(V_{RF}^- - V_{LO,DC} - V_{Tn,4} - \frac{V_{LO}^+ - V_{LO,DC}}{2} \right) \cdot \left(V_{LO}^+ - V_{LO,DC} \right)
\end{aligned}
\tag{6.1}
$$

The bulk effect gives in first order a linear change around the bias point which is canceled with the double balanced structure [Song JSSC86]. Assuming perfect matching and no bulk-effect, the output signal is then :

$$
\begin{aligned}
V_{out}^+ - V_{out}^- &= R_f \cdot ((I_1 - I_4) - (I_3 - I_2)) \\
&= \beta \cdot R_f \cdot (V_{RF}^+ - V_{RF}^-) \cdot (V_{LO}^+ - V_{LO}^-)
\end{aligned}
\tag{6.2}
$$

Mismatch between the input transistors has two effects. β and V_T mismatches both result in the appearance of residual DC-offset voltages. These offset voltages either appear directly on the output of the mixer or they result in direct feedthrough of the RF and LO signal to the output caused by multiplication of these signals with the offset voltage. The RF and LO signal are however high frequency signals. A direct feedthrough of these signals will therefore be strongly suppressed by the added capacitors C_v. The second effect, caused by β mismatch, is the appearance of a quadratic V_{LO} component in the output signal (see section 2.4.2 for more information on this effect).

$$\begin{aligned} V_{out}^{+} - V_{out}^{-} &= \beta \cdot R_f \cdot (V_{RF}^{+} - V_{RF}^{-}) \cdot (V_{LO}^{+} - V_{LO}^{-}) \\ &\quad + \Delta\beta \cdot R_f \cdot (V_{LO}^{+} - V_{LO}^{-})^2 \end{aligned} \tag{6.3}$$

This explains why the RF signal is best applied to the gates of the modulating transistors. A quadratic V_{RF} component would be highly unwanted. The squared RF signal has frequency components at twice its center frequency and at the baseband. The high frequency components are filtered out, but the baseband components will degrade the wanted baseband signal. The bandwidth of this parasitic baseband signal is equal to the bandwidth of the RF signal (e.g. 100 MHz). Most of its power will however be situated at the lower frequencies, in a band equal to the correlation bandwidth of the RF signal (which is about equal to the bandwidth of a transmission channel, e.g. 200 kHz). The squared LO signal results only in an extra DC component at the output of the opamp. This DC signal can also be a problem. It can be as large as the wanted signal and in that case it will saturate the succeeding filters, but with the use of dynamic DC suppression techniques these components can be sufficiently suppressed without generating too much distortion [Rab ACD93] (see also section 2.4.2).

6.2.2.1 Mixer design. The DC biasing levels of the RF and LO signal must be chosen carefully. There will be a lot of distortion when the modulating transistors are not kept in the triode region at all times. Fig. 6.2 and equation 6.4 define the voltages applied to the sources, drains and gates of the modulated transistors. $m(t)$ is in this equation the normalized antenna signal, varying between -1 and $+1$.

$$\begin{aligned} V_{RF}^{+} &= V_{RF,DC} + V_{RF,AC} \cdot m(t) \\ V_{RF}^{-} &= V_{RF,DC} - V_{RF,AC} \cdot m(t) \\ V_{LO}^{+} &= V_{LO,DC} + V_{LO,AC} \cdot \sin(\omega_0 t) \\ V_{LO}^{-} &= V_{LO,DC} - V_{LO,AC} \cdot \sin(\omega_0 t) \end{aligned} \tag{6.4}$$

The smallest possible level that can appear at the gates is $V_{RF,DC} - V_{RF,AC}$. This level must be at least a V_T higher than the largest possible source level, which is $V_{LO,DC}$. Otherwise the transistors will be turned off during a part of each mixing period. Saturation of the modulated transistors will appear when the largest possible drain-source voltage V_{DS} becomes higher than the smallest saturation voltage $V_{DS,sat}$ that can appear. The highest possible drain-source voltage is $V_{LO,AC}$. The saturation

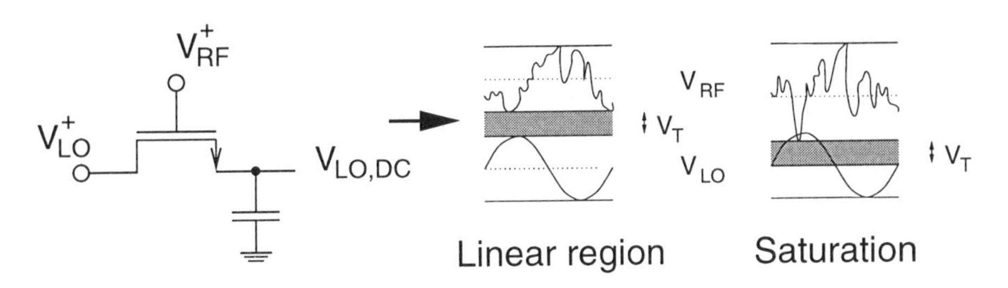

Figure 6.2. Definition of the different voltages applied to the source, drain and gate of one of the mixing transistors.

voltage is equal to $V_{GS} - V_T$ and its smallest possible level is $V_{RF,DC} - V_{RF,AC} - V_{LO,DC} - V_T$. Continuous operation in the linear region is thus guaranteed when :

$$V_{GS,DC} > V_{RF,AC} + V_{LO,AC} + V_T \tag{6.5}$$

Continuous operation in the linear region is not necessary. Saturation does not directly result in distortion. Operation in the saturation region results in the generation of quadratic components, but the cross-coupled double balanced structure makes sure that all quadratic components in the voltage to current conversion characteristic of the modulated transistors are cancelled out [Baban JSSC85, Song JSSC90]. It is only under the influence of mismatch between the four mixing transistors that they will interfere with the correct operation of the downconversion mixer. Therefore, excursion in the saturation region may be allowed up to a certain point. The lower limit for correct operation becomes :

$$V_{GS,DC} > V_{RF,AC} + V_T \tag{6.6}$$

The proposed downconversion mixer is designed to operate on a 5 V power supply voltage in a standard 1.2 μm CMOS process. Information on the proposed process can be found in appendix A. The LO DC level is taken to be 1.15 V, making the maximal LO signal that can be applied 4.6 V_{ptp} differential (this is equivalent to 17.2 dBm). An RF DC level of 3.85 V allows an input signal of 4 times $V_{RF,DC} - V_{LO,DC} - V_T$, which gives 8 V_{ptp} differential or 22.0 dBm for the 1.2 μm CMOS process. These are very high values for an RF mixer, resulting in a topology which renders an excellent linearity. Saturation will occur when $4 \cdot (V_{RF,AC} + V_{LO,AC}) > 8\ V_{ptp}$. The applied LO signal is for this reason limited to 2.5 V_{ptp} differential, so that an RF signal of 5.5 V_{ptp} still can be applied without driving any of the modulated transistors into its saturation region.

The use of a large $V_{GS} - V_T$ makes it possible to realize a large nominal (i.e. biased at $V_{RF,DC}$) conductance g_{DS} with a small transistor, which gives small parasitic capacitances (less than 20 fF) and thus a high input bandwidth. The W/L of the modulated transistors is 6, their g_{DS} is 1 mS. The four modulated transistors are the only high frequency part on the chip and they take up very little area. Because of their small parasitic capacitances the on-chip RF-to-LO crosstalk, which is mainly determined by the absolute mismatch between C_{GS} values, is also very small. Hence, RF to LO crosstalk is mainly caused by off-chip crosstalk and crosstalk between the bonding wires. The mismatch between small transistors may be large, resulting in a larger self-mixing of the LO signal. A trade-off has to be made here, but in general more LO self-mixing can be allowed than RF-to-LO crosstalk (see previous section).

The use of a large $V_{GS} - V_T$, and thus small mixing transistors, has also as effect that the input bandwidth is completely determined by capacitances of the bonding pads. As a result very high frequency performances can be achieved.

6.2.2.2 Frequency Domain Behavior. The well known negative feedback configuration of fig. 6.3 suppresses any signal at the virtual ground node when a perfect opamp is used. An opamp has however always a limited gain-bandwidth GBW and DC-gain A_0. With a limited GBW the transfer function from the input to the virtual ground becomes (A_0, C_f and C_v are not taken into account) :

$$\frac{V_v}{V_{in}} = \frac{\frac{R_f}{R_{in}} \cdot \mathrm{j}\frac{\omega}{2\pi \cdot GBW}}{1 + \frac{R_f + R_{in}}{R_{in}} \cdot \mathrm{j}\frac{\omega}{2\pi \cdot GBW}} \approx \frac{A \cdot \mathrm{j}\frac{\omega}{2\pi \cdot GBW}}{1 + A \cdot \mathrm{j}\frac{\omega}{2\pi \cdot GBW}} \tag{6.7}$$

The transfer function from the input to the output is :

$$\frac{V_{out}}{V_{in}} = \frac{-\frac{R_f}{R_{in}}}{1 + \frac{R_f + R_{in}}{R_{in}} \cdot \mathrm{j}\frac{\omega}{2\pi \cdot GBW}} \approx \frac{-A}{1 + A \cdot \mathrm{j}\frac{\omega}{2\pi \cdot GBW}} \tag{6.8}$$

Normally, the conclusion to be drawn from these equations is that the GBW must be a certain factor higher than the highest frequency component that has to be processed. This is also true if the circuit would be used as lowpass filter with an extra capacitor C_f in the feedback loop. However, here in the mixer topology the situation is different. There is a very large spectrum of input currents to the virtual ground node, basically starting from DC up to twice the LO frequency (more than 2 GHz), but the wanted signal takes up only a few hundred kHz situated at the baseband. The output bandwidth BW, given in equation 6.8 as GBW/A, only has to be 500 kHz or more. By designing it to be 1 MHz this specification is fulfilled independent of absolute parameter variations. Equation 6.8 shows that this limited bandwidth can be realized by using an opamp with a small GBW. In combination with a feedback capacitor C_f

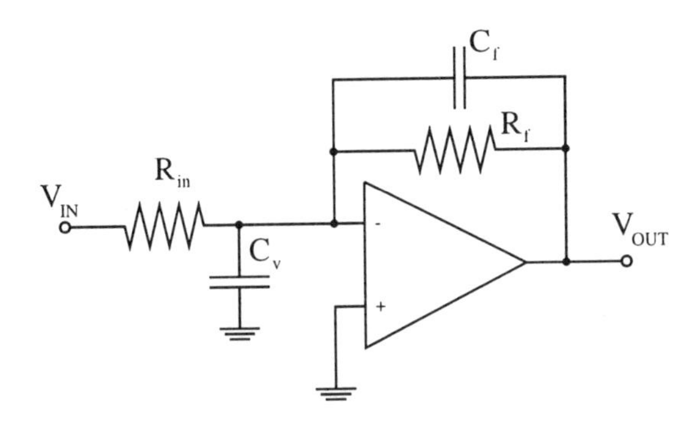

Figure 6.3. The negative feedback configuration (represented single-ended).

the bandwidth can be lowered further and positioned more accurately. The use of an opamp with a small GBW poses however a problem. As equation 6.7 reflects, beyond the BW the input signal is directly transferred to the virtual ground node. It is not suppressed anymore by the opamp via the feedback construction. The solution for this problem can be found in adding extra capacitance to the virtual ground nodes. This introduces an extra pole in both equation 6.7 and 6.8 situated at $1/(2\pi \cdot R_{in}C_v)$. In order to suppress the signals appearing on the virtual ground without changing the output bandwidth, its value must be positioned between $1/(2\pi \cdot R_f C_f)$ and the BW. Fig. 6.4 displays the transfer function from the input to the virtual ground and the output, asuming a perfect opamp with an open-loop gain A_0.

6.2.2.3 Opamp Design. The opamp topology is shown in fig. 6.5. It is a symmetrical folded cascode structure succeeded by a source follower which performs buffering and level shifting. The opamp runs on a 5 V power supply voltage. The output DC level is 1.15 V. It is kept on this level with a standard type common-mode feedback : the common-mode signal of the output is measured at the middle of a high-ohmic resistor placed between the two output nodes. Table 6.1 gives the W and L values for the different transistors in the proposed opamp. The GBW of the opamp is 100 MHz, but because of the partial feedback the second, non-dominant, pole can be situated at only 30 MHz. The feedback resistors over the opamp are 60 kΩ, which makes the conversion gain very high. The conversion gain is almost 20 dB for a 12 dBm LO signal. An extra lowpass filter of 1 MHz is implemented with the feedback capacitors C_f (2.5 pF). This small output bandwidth and the high conversion gain are required by for instance a zero-IF or low-IF type receiver structure. The extra filtering and amplification that is performed in the output stage of the mixer relaxes in these topologies the specifications for both the output stage of the opamp and the succeeding stages of the receiver because of the reduced dynamic range of the output signal when it is more filtered.

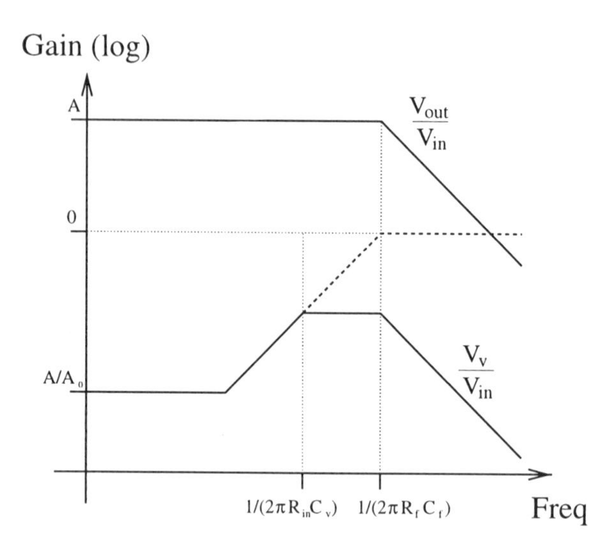

Figure 6.4. Transfer function of the mixer used as amplifier (the dotted line gives the transfer function to the virtual ground node when C_v is not present).

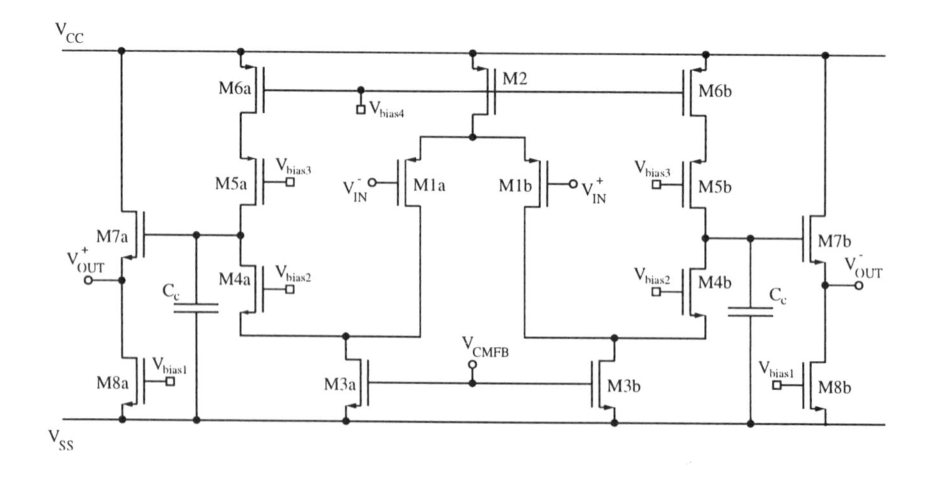

Figure 6.5. The circuit topology of the low frequency opamp.

Table 6.1. Device sizes for the low frequency opamp.

DEVICE	TYPE	SIZES
M1	pMOS	$w = 864\ \mu$m, $l = 1.5\ \mu$m
M2	pMOS	$w = 432\ \mu$m, $l = 1.5\ \mu$m
M3	nMOS	$w = 168\ \mu$m, $l = 1.2\ \mu$m
M4	nMOS	$w = 552\ \mu$m, $l = 3.0\ \mu$m
M5	pMOS	$w = 864\ \mu$m, $l = 3.0\ \mu$m
M6	pMOS	$w = 312\ \mu$m, $l = 2.2\ \mu$m
M7	nMOS	$w = 42\ \mu$m, $l = 1.2\ \mu$m
M8	nMOS	$w = 312\ \mu$m, $l = 1.2\ \mu$m
Cc	Cap	$C = 3$ pF

6.2.2.4 Noise, Power Consumption and Optimization. The noise figure (NF) of the HF mixer is a very important parameter in receiver design. In a receiver the mixer is positioned directly after the LNA and a mixer with a high NF can only be used in combination with a high quality LNA. In this case the LNA must have a high gain which, in turn, can only be allowed when the antenna signal is first filtered with a high Q HF filter. The problem with mixers is that their NF can not be made arbitrarily small. Their NF can not even be in the proximity of the NF's that can be achieved with LNA's. This is due to the intrinsic nature of the mixing process.

The two main noise sources in the presented mixer are the modulated input transistors in the triode region and the input stage of the LF opamp. The thermal output noise density generated by these devices is (with the factor 2 introduced by the differential operation) :

$$dv_{out}^2 = 2 \cdot 4kT \cdot \left(\frac{2}{3} \cdot \frac{1}{g_{m,in,\text{eq}}} \cdot (R_f \cdot 2g_{DS})^2 + 2g_{DS} \cdot R_f^2 \right) \cdot df \qquad (6.9)$$

Equation 6.9 gives the output noise density for the mixer biased in its static operation point. The modulated transistors have however, under transient conditions, always the same impedance ($2g_{DS}$) to the virtual ground nodes. For this reason equation 6.9 is, for the presented topology, also a good measure for the output noise density under transient operation conditions. The equivalent input noise can be found by dividing this expression by the conversion gain G. The conversion gain is defined as the ratio of the output signal over the input signal ($V_{out}^+ - V_{out}^- / V_{RF}^+ - V_{RF}^-$) and follows from equation 6.2 :

$$
\begin{aligned}
G &= \frac{(V_{out}^{+} - V_{out}^{-})_{\mathrm{ptp}}}{(V_{RF}^{+} - V_{RF}^{-})_{\mathrm{ptp}}} \\
&= \frac{1}{\sqrt{2}} \cdot \frac{1}{\sqrt{2}} \cdot \beta \cdot R_f \cdot (V_{LO}^{+} - V_{LO}^{-})_{\mathrm{ptp}}
\end{aligned} \tag{6.10}
$$

The first $\sqrt{2}$ appears because the multiplication is performed with a sine and not a square wave. This implies that the RMS value of the LO signal has to be used. The second $\sqrt{2}$ is necessary because only the low frequency mixing product is regarded as wanted. Dividing equation 6.9 by equation 6.9 results in :

$$
\begin{aligned}
\mathrm{d}v_{in}^2 &= \frac{1}{G^2} \cdot \mathrm{d}v_{out}^2 = \left(\frac{2}{R_f \cdot \beta \cdot (V_{LO}^{+} - V_{LO}^{-})_{\mathrm{ptp}}} \right)^2 \cdot \mathrm{d}v_{out}^2 \\
&= 8kT \cdot \left(\frac{4 g_{DS}}{\beta \cdot (V_{LO}^{+} - V_{LO}^{-})_{\mathrm{ptp}}} \right)^2 \cdot \left(\frac{2}{3} \cdot \frac{1}{g_{m,in,\mathrm{eq}}} + \frac{1}{2 g_{DS}} \right) \cdot \mathrm{d}f \\
&\doteq 8kT \cdot \left(2 \cdot \frac{2 g_{DS}}{\Delta g_{DS}} \right)^2 \cdot \left(\frac{2}{3} \cdot \frac{1}{g_{m,in,\mathrm{eq}}} + \frac{1}{2 g_{DS}} \right) \cdot \mathrm{d}f
\end{aligned} \tag{6.11}
$$

The factor $4g_{DS}/\Delta g_{DS}$ is the extra term intrinsic to any mixer or double balanced structure (see section 4.2.2.2). It is equal to the ratio between the gain of the mixer used as single balanced amplifier and the conversion gain A/G. For low noise, this term should be as low as possible. The noise can also be lowered by using a larger $g_{m,in,\mathrm{eq}}$ and g_{DS}, but this is at the expense of a higher power consumption. The minimal value for this extra term is found when the swing of the LO signal is maximal.

$$
\begin{aligned}
\frac{A}{G} &= 2 \cdot \frac{2 g_{DS}}{\Delta g_{DS}} = \frac{4 \cdot \beta \cdot (V_{GS} - V_T)}{\beta \cdot (V_{LO}^{+} - V_{LO}^{-})_{\mathrm{ptp}}} \\
&= 4 \cdot \frac{V_{RF,DC} - V_{LO,DC} - V_T}{(V_{LO}^{+} - V_{LO}^{-})_{\mathrm{ptp}}}
\end{aligned} \tag{6.12}
$$

$$
\left(\frac{A}{G} \right)_{\mathrm{min},RF@gates} = \frac{V_{RF,DC} - V_{LO,DC} - V_T}{2 \cdot V_{LO,DC}} \tag{6.13}
$$

$$
\left(\frac{A}{G} \right)_{\mathrm{min},LO@gates} = \frac{V_{RF,DC} - V_{LO,DC} - V_T}{2 \cdot (V_{RF,DC} - V_{LO,DC} - V_T)} = 2 \tag{6.14}
$$

So, according to equation 6.14 for the case in which the LO signal is applied to the gates of the modulated transistors, this means that the mixing function adds at least 6 dB to the NF. For the rest the NF is directly determined by the conductivity of a single modulated transistor (g_{DS}) and by the transconductance value of the opamp input stage ($g_{m,in,\mathrm{eq}}$), just like in any other amplifier with negative feedback. The noise of the opamp input stage can be made sufficiently small without requiring too much power by using large input transistors (and a small $V_{GS} - V_T$). This is possible because the opamp only has to be LF and there is already standing 25 pF (C_v) at the input nodes. The only limitation is that the internal slew-rate of the output may become too low when its input transistors are too large for a given transconductance. The value of g_{DS} can not be made arbitrarily large because this conductor has to be driven by the LO, resulting for a high g_{DS} of the mixing transistors in a high power consumption for the LO signal generator.

From equation 6.13 it might seem that the NF can be made arbitrarily small when the RF signal is applied to the gates by taking $V_{RF,DC} - V_T < 2 \cdot V_{LO,DC}$. This implies however a reduction of the input swing and in this way the dynamic range (DR) at the input is not improved. It is not only necessary to minimize the NF of a mixer. It is its input signal capability (i.e. the DR) which must be optimized and compared with the power consumption. A good measure for the performance of a mixer is its efficiency as defined in section 4.2.2.2. Here this gives :

$$\begin{aligned}
\eta_{RF@gates} &= \frac{DR}{P} \cdot kT \cdot BW \\
&= \left(\frac{S}{N}\right)^2 \cdot \frac{1}{P} \cdot kT \cdot BW \\
&= \frac{(V_{RF,DC} - V_{LO,DC} - V_T)^2}{8kT \cdot \left(\frac{A}{G}\right)^2 \cdot \frac{1}{2g_{DS}} \cdot BW} \cdot \frac{1}{(V_{LO}^{+} - V_{LO}^{-})^2 \cdot g_{DS}} \cdot kT \cdot BW \\
&= \frac{1}{64} \approx 1.5\ \% \qquad (6.15)
\end{aligned}$$

The power efficiency of a high quality low frequency amplifier is about 10 %. The power efficiency of the presented mixer topology is 1.5 %. This value takes the power efficiency of the LO signal source not into account. 1.5 % is a very high value for high frequency mixers, which have intrinsically a low efficiency (for comparison : a typical Gilbert mixer with a 12 dB NF, a +6 dBm IP3 and a power consumption of 3 mW, has an efficiency of 0.2 % [Gilb JSSC68]). Only a commutating mixer with a downconversion topology can give a better efficiency. Theoretically its efficiency can be up to 1.4 times better than the efficiency of a true sinusoidal mixer as presented here.

Equation 6.15 is slightly different when the LO signal is applied to the gates of the modulated transistors instead of the RF signal :

$$\begin{aligned} \eta_{LO@gates} &= \left(\frac{S}{N}\right)^2 \cdot \frac{1}{P} \cdot kT \cdot BW \\ &= \frac{V_{LO,DC}^2}{8 \cdot \left(\frac{A}{G}\right)^2 \cdot \frac{1}{2g_{DS}}} \cdot \frac{1}{V_{LO,DC}^2 \cdot g_{DS}} \end{aligned} \tag{6.16}$$

And only when A/G is minimal this reduces to :

$$\eta_{LO@gates,\max} = \frac{1}{64} \approx 1.5\ \% \tag{6.17}$$

The main difference is that in this case the efficiency of 1.5 % is only obtained with the maximal LO signal. This is the case for most mixer topologies. Commutating mixers have an LO signal whose fundamental component is larger than the maximum applicable LO signal swing, explaining their 1.4 times higher efficiency. The efficiency of 1.5 % is in the situation of equation 6.15 always obtained, independent of the amplitude of the LO signal. This is caused by the fact that, when the LO signal is applied to the sources of the modulated transistors, a lower conversion gain is compensated by a lower power consumption for the LO signal source. This is another reason why the topology with the RF signal applied to the gates is preferred over the topology in which the LO signal is applied to the gates of the modulated transistors.

6.2.2.5 Realization and Measurement Results. The mixer has been designed and realized in a 1.2 μm CMOS process [Crols ESSC94]. Main design goals where to illustrate the high linearity and high input frequency capabilities of the proposed topology. The RF and LO signal are externally made differential by means of baluns. They are, as stated in section 6.2.2.1, chosen to be biased at respectively 3.85 V (for the RF signal, applied at the gates) and 1.15 V (for the LO signal, applied at the sources and drains). The measurement setup with the baluns is shown in fig. 6.6. Fig. 6.7 shows the measured input bandwidth. It is from DC to 1.5 GHz. The input bandwidth is measured at 6 dB attenuation, because both input signals are falling off at the same time. The measured 1.5 GHz maximum input bandwidth is equal to the specified bandwidth of the baluns which have been used to generate the differential input signals. The output bandwidth is determined by the untuned value of $1/R_f C_f$. For the application in mind (zero-IF and low-IF receivers) it must be at least 500 kHz and it is designed to be 1 MHz so that it can cope with the large absolute processing value spreads. The measured output bandwidth is 780 kHz. The measured conversion gain is 18 dB for a 12 dBm differential LO signal.

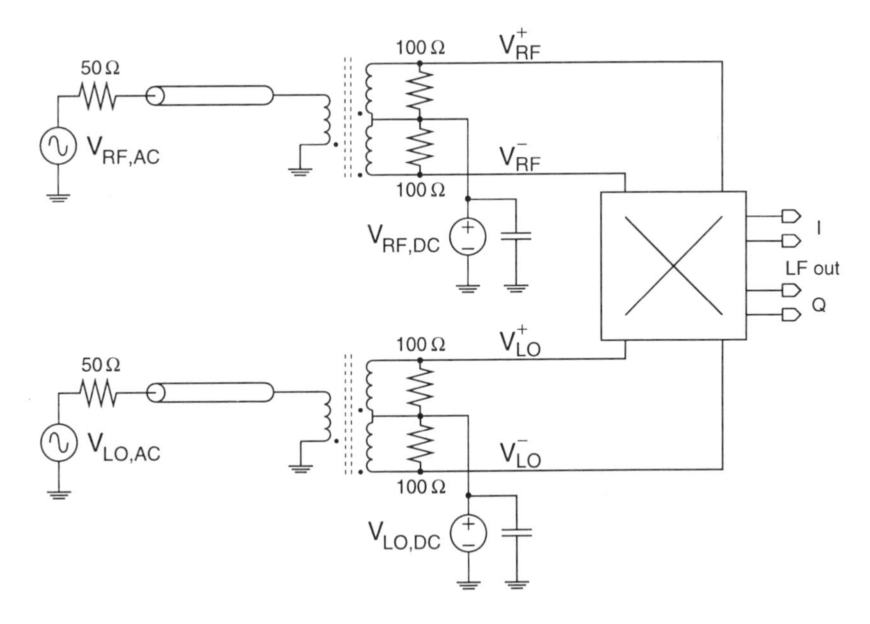

Figure 6.6. Measurement setup for the highly linear downconversion mixer.

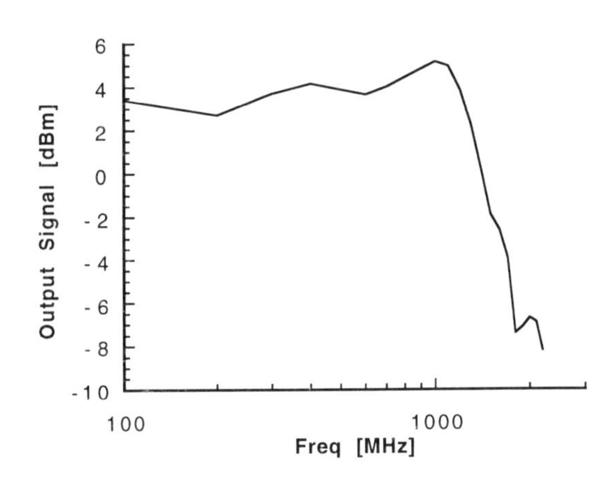

Figure 6.7. The measured input bandwidth.

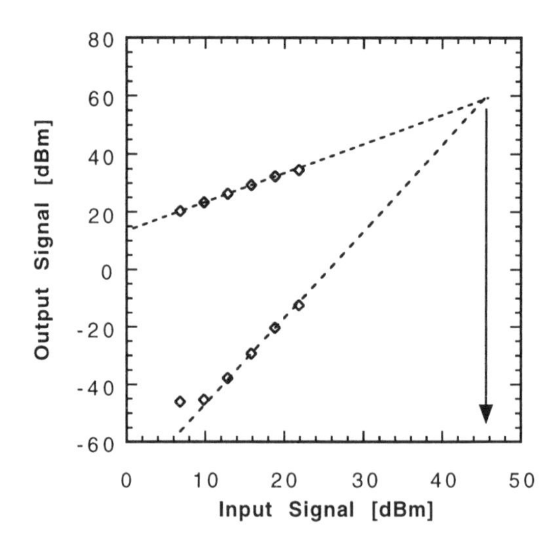

Figure 6.8. The IP3 measurement.

The IP3 measurement, shown in fig. 6.8, proves that the linearity is very high. The IP3 is 45.2 dBm. This value does not depend on the magnitude of the LO signal. An input signal of 22 dBm, the theoretical value for the maximal applicable input swing, gives an IM3 of 46.4 dB. The IM3 measurement was performed with the downconverted carriers lying outside of the output band (at 10 MHz) and the IM3 product lying in the passband (at 200 kHz). This made it possible to measure the distortion of the input stage and not the output stage for signals which would normally produce an output signal of up to 40 dBm. The measured noise figure (NF) is 32 dB. This is mainly caused by the input stage of the opamp which has not been optimized for low noise. By using large input transistors for the low frequency opamp, which has no effect on the high frequency performance of the mixer, the noise figure can easily be improved down to 24 dB. The mixer is designed in a 1.2 μm CMOS process. The measured power consumption is 1.3 mW which is all taken up in the low frequency opamp. The total chip area is 1 mm^2. The modulated transistors only take up 300 μm^2 of this area. A microphotograph of the realized chip is shown in fig. 6.9.

6.2.3 Upconversion Mixers in CMOS

From a high level point of view it may seem that the same function has to be realized for upconversion and downconversion. After all, both require the implementation of a four-quadrant multiplier. From a practical design point of view this is of course

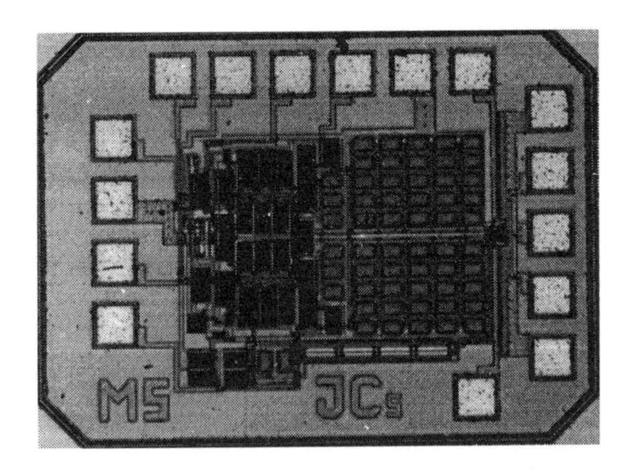

Figure 6.9. Microphotograph of the highly linear CMOS downconversion mixer.

not true. In the previous sections it was for instance illustrated how the differences between up- and downconversion can be exploited in order to achieve the highest possible performance for a downconversion mixer.

The main difference between a downconversion mixer and an upconversion mixer is that the upconversion mixer requires a high frequency output. The multiplication itself is often performed by modulating the transconductance properties of a cross-coupled structure of transistor devices. Cross-coupling is required to make the multiplier a four-quadrants multiplier. In the Gilbert multiplier the (non-linear) transconductance of four cross-coupled bipolar transistors is modulated by varying its bias currents [Gilb JSSC68, Gilb ISSCC83]. Fig. 6.10a shows the principles of this configuration. Fig. 6.10b shows that the same principle can be applied to the four transistor MOS structure of [Song JSSC86]. The transconductance is here modulated by varying the drain-source voltages V_{DS} of these transistors. In fig. 6.10b the MOS devices are operating in the linear region which gives a linear dependence of the transconductance on its modulating voltage. In this way improved linearity results can be obtained when using a CMOS technology instead of a silicon bipolar technology [Song JSSC86, Crols JSSC95a].

The drawback of the use of a CMOS technology is its lower frequency performance. This is not so much a problem of the modulation process itself [King CICC96, Crols JSSC95a]. In present day sub-micron technologies nMOS devices can achieve an f_{max} (the maximal bandwidth of a transistor, defined as $g_m/2\pi(C_{gs} + C_{ds})$ [Stey ACD93]) of well over 2 GHz. The mixing frequencies that can be achieved are mainly determined by this value and modulation of sub-micron nMOS devices at a frequency of 1 GHz is thus possible.

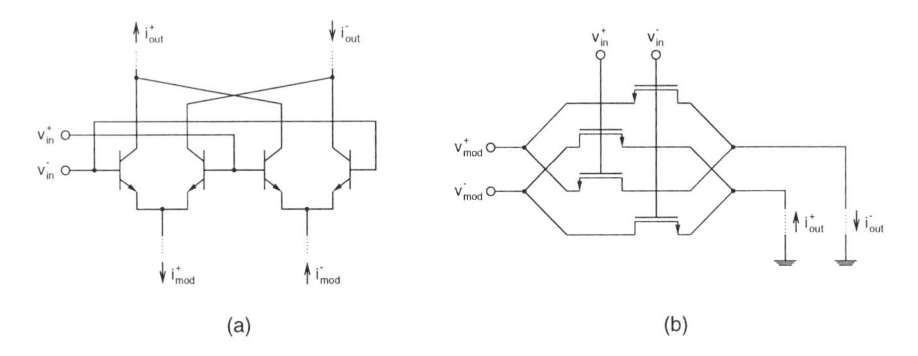

Figure 6.10. Basic schematic of the mixing process performed in a) a bipolar Gilbert multiplier and b) in a CMOS multiplier biased in the linear region.

The problem for the realization of a CMOS upconversion mixer at 1 GHz is the output current generated during the modulation process. This current has to be buffered and amplified. For a downconversion mixer this is no problem because there the output current is a low frequency signal and this property can be used to realize a high quality CMOS downconverter (see section 6.2.2). The realization of an upconversion mixer for a transmitter requires however the realization of a transimpedance amplifier (i.e. a current to voltage converter) which often not only has to be capable of achieving a 1 GHz bandwidth with a good linearity, but also has to drive a high enough signal into an off-chip 50 Ω load. Gain implies the use of devices with an f_{max} that is at least as large as the product of the required bandwidth and the required gain. A high linearity can only be achieved with some kind of feed-back mechanism [Wamba KUL96] and this implies the realization of an open-loop gain for the transimpedance amplifier that is even higher than the required closed-loop gain. The consequence is that transistor devices with an f_{max} of up to 10 GHz and even more are required. These required CMOS device performances are only achieved in deep sub-micron processes. In these technologies the realization of high performance CMOS upconverters becomes possible [King CICC96, King KUL96, Rofou CICC96, Crols ESSCC96].

6.3 SPIRAL INDUCTORS

6.3.1 CMOS Output Buffers

The realization of CMOS buffers which can drive a 50 Ω load with a good linearity at 1 GHz is one of the main problems for the realization of a full CMOS transceiver at 1 GHz. Spiral inductors can be used to significantly improve the design of this building block. Fig. 6.11 shows the most basic amplifier stage (a amplifier,resistor loaded common source amplifier) and its amplifier,inductor loaded equivalent. A first-order calculation of the f_{max} requirements for the two transistors demonstrates how the design can be improved by means of inductors. Both amplifiers have to realize a given gain A $(= g_m \cdot R_L)$ at respectively a given corner or center frequency f_c (which would be around 1 GHz). For the resistor loaded amplifier is :

$$f_c = \frac{1}{2\pi R_L(C_L + C_{ds} + C_{gd})} \tag{6.18}$$

For high frequency design, transistors with minimal gate length L will be used. Transistors with a constant L have a constant ratio between their parasitic capacitances which only depends on the technology. For the 0.7 μm CMOS process of appendix A, α of equation 6.19 is for instance approximately equal to 2.2 for an nMOS transistor operating in saturation.

$$f_c = \frac{1}{2\pi R_L C_L + 2\pi \cdot \frac{A}{g_m} \cdot \frac{C_{gs} + C_{ds}}{\alpha}} \tag{6.19}$$

This gives the frequency requirements for the nMOS transistor of the resistor loaded amplifier.

$$f_{max,R} = \frac{\alpha}{A} \cdot \frac{1}{\frac{1}{f_c} - 2\pi R_L C_L} \tag{6.20}$$

The same calculation can be made for the inductor loaded amplifier, but in that case equation 6.18 is the expression for the bandwidth BW $(= f_c/Q)$. The parasitic capacitance of the spiral inductor C_P must be added to the value of C_L.

$$f_{max,L} = \frac{\alpha}{A} \cdot \frac{1}{\frac{Q}{f_c} - 2\pi R_L \cdot (C_L + C_P)} \tag{6.21}$$

A comparison between equation 6.20 and equation 6.21 shows that it is beneficiary to use inductors with a low parasitic capacitance when they are available. In fact, the use of an inductor loaded amplifier instead of a resistor loaded amplifier is only beneficiary under the following condition :

$$\frac{f_c}{Q - 1} < \frac{1}{2\pi R_L C_P} \tag{6.22}$$

Apart from the fact that high quality inductors must be available, there are two other conditions that must be fulfilled. First of all, bandpass operation must be allowed. This is however the case in the high frequency part of wireless transceivers. Secondly, it must be possible to control the center frequency of the amplifier very accurately.

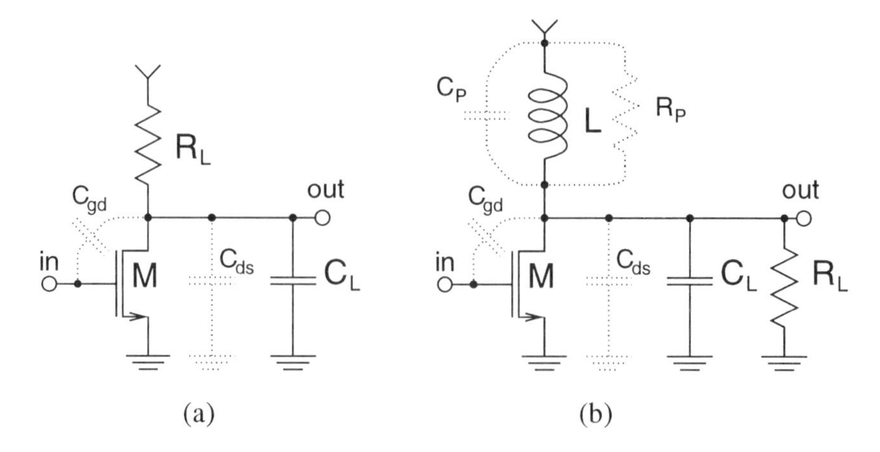

Figure 6.11. Comparison between a) a resistor loaded amplifier and b) an inductor loaded amplifier (parasitics are in dotted lines).

If not, only a low Q can be used in order to assure that the operation frequency will always be situated within the amplifier's passband. In practice, this will mean that with on-chip inductors only a Q between 1 and 3 is allowed.

Another conclusion that can be drawn from equation 6.22 is that inductors can only be used when R_L is low (e.g. 50 Ω). This means that using inductor loading on internal nodes within a chip is not beneficiary, because on these nodes larger impedance values (150 to 300 Ω) are normally used. These larger values do not only lower the value of the $1/(2\pi R_L C_P)$ term via the higher value of R_L. A higher impedance R_L will also demand that the parasitic parallel resistance of the inductor R_P is larger and this can only be achieved by using thicker conductors for the inductor. This means a larger inductor geometry and consequently also a higher value for C_P.

6.3.2 On-Chip Inductors

The design of RF building blocks can be drastically improved when also inductors are used [Cran VLSI96, Burgh IEDM95]. With passive inductors, bandpass operations can be made with a significantly lower power consumption. This is not the case for active bandpass operations which use simulated inductors. In section 4.2.2.2 it was already illustrated that the power consumption of a passive building block is proportional to its bandwidth BW. A lowpass building block that has to process a bandpass signal at 900 MHz will require a bandwidth of at least 900 MHz, while a bandpass building block will only need a bandwidth of e.g. 100 MHz to process this signal. Other advantages of the use of inductors are the higher operating frequencies that can be achieved (see the previous section) and the lower power supply voltages that can be

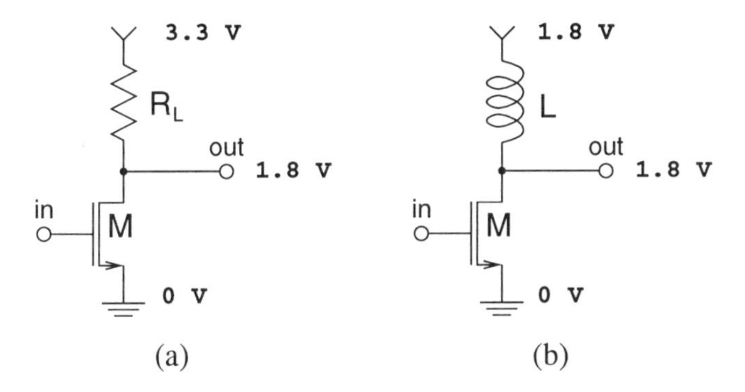

Figure 6.12. Lower power supplies can be used when designing with inductors.

used. Fig. 6.12 illustrates why lower power supply voltages can be used when designing with inductors. On-chip signals will however still be as high as without inductors.

The use of off-chip inductors has only limited advantages. Parasitic capacitances of off-chip components and interconnections are much higher than on-chip. This reduces the advantage of being able to work at higher frequencies. Lower impedances are needed in order to achieve the same bandwidth, resulting in a higher power consumption.

The alternative is the use of on-chip spiral planar inductors. Two main problems arise when using planar inductors on silicon substrates. First of all, there are the large resistive and capacitive parasitics originating from losses in the aluminum conductors, losses in the conducting substrate and capacitive coupling to this substrate. In GaAs, where gold is used for conductors and where the substrate is non-conductive, these parasitics are far less important and planar inductors are widely used [Haigh 1989, Kim IEDM95]. In silicon, the substrate loss and coupling can be eliminated by etching away, with an extra processing step, the substrate under the inductor, either via front [Chang IEDL93, Rofou ISSCC96] or back etching [Based ESSC94]. Fig. 6.13 shows an example of such etched inductors. Etching requires however an extra processing step, which is costly, reduces yield and reliability and which is not available in any standard silicon process.

A second problem, which occurs for silicon and GaAs, is how to design circuits with planar inductors. Only two methods are available : either one uses planar inductors with a geometry which has already been implemented and measured on a previous run of the process [Nguy JSSC90], or one uses time consuming electro-magnetic simulations to determine the characteristics of a chosen geometry [Chang IEDL93]. In both cases, the performance of a circuit is determined by the fixed inductor geometry, without leaving any room to gain an insight in the trade-offs between specifications.

(a)

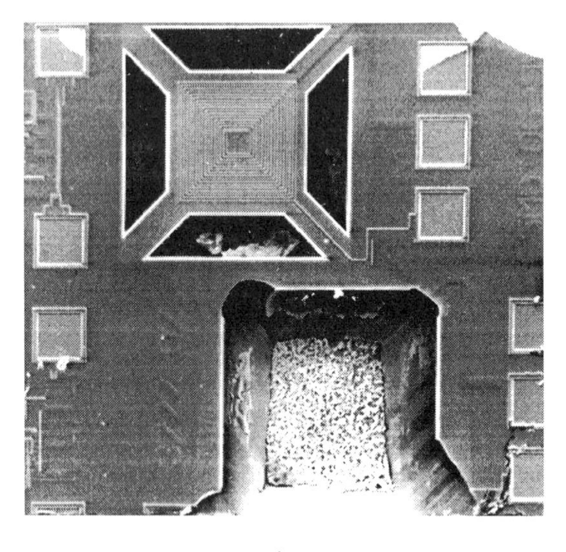

(b)

Figure 6.13. Example of a planar inductor on a silicon substrate under which the substrate is etched away with an extra processing step.

A structured analog design methodology requires the opposite approach where the optimal inductor geometry is determined by the required circuit specifications.

In this text a model for planar inductors and their parasitics on a lowly doped silicon substrate is introduced. The behavior of inductors on lowly and heavily doped substrates is compared and it is shown that the electro-magnetic losses due to the substrate are negligible for processes which use lowly doped substrates. Many processes and

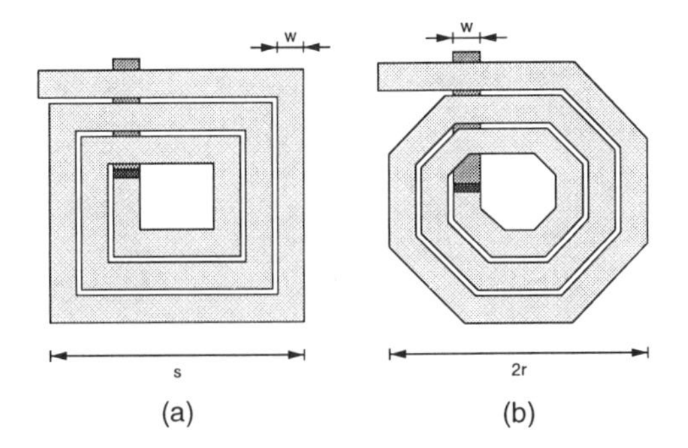

Figure 6.14. Layout of a) a square and b) a near circular spiral inductor.

even newly developed deep sub-micron standard CMOS processes use lowly doped substrates, creating in this way great opportunities to use planar inductors in standard silicon processes without requiring any special techniques like substrate etching.

6.3.3 Model for Planar Inductors

Fig. 6.14 shows two types of layout for planar spiral inductors. Fig. 6.15 gives a lumped model for these planar inductors and table 6.2 gives the analytical expressions of the proposed model for the inductance and the parasitics of such planar inductors [Crols VLSI96]. Table 6.3 gives an overview and description of the parameters used in this model.

The inductance L does not depend on the technology, only on the geometry of the inductor. The proposed expression models the inductance with an accuracy better than 10 % for a wide range of different geometries. A less accurate analytical expression for the inductance can be found in [Voorm 1993]. The model presented in this text takes into account such parameters as the spacing between the metal lines (via η_{Ar} and η_w) and the extend of the open area in the middle of the inductor (via η_{Ar}). An accuracy much better than 10 % can not be obtained as the proposed model is a continuous representation of a device with a discrete parameter (the number of turns). This is however only true for inductors with a low number of turns (3 to 4 turns). Inductors with more turns could be modeled more accurately, but the use of one model for both square and near circular spiral inductors also limits the achievable accuracy. The proposed expression for the inductance holds only for inductors in free air or on lowly doped substrates. Highly doped substrates have a high conductivity, allowing the generation of eddy-currents in the substrate. This results in a reduction of the magnetic field and thus of the effective inductance value [Cran VLSI96]. On lowly doped substrates, the inductance reduction is below 1 %.

Table 6.2. Overview of the equations for the proposed analytical model for planar spiral inductors.

MODEL EQUATIONS
$L = K_L \cdot \frac{Ar^{3/2}}{w^2} \cdot \eta_{Ar}^{5/3} \cdot \eta_w^{\alpha}$ with $\alpha = \begin{cases} 1/4, & \text{for square inductors} \\ 1/7, & \text{for circular inductors} \end{cases}$
$R_S = \frac{\rho_{metal}}{t_{metal}} \cdot \frac{Ar}{w} \cdot \left(\frac{1}{w} + \frac{1}{\delta}\right) \cdot \eta_{Ar}$ with $\delta = \left(\frac{\rho_{metal}}{t_{metal}} \cdot \frac{K_R}{\omega_0}\right)^2$
$C_P = \frac{\varepsilon_{ox}}{t_{ox}} \cdot Ar \cdot \eta_{Ar}$
$R_C \approx R_{\square,sub} \cdot \left(\frac{t_{sub}^2}{t_{sub}^2 + Ar}\right)$

Table 6.3. Overview of the parameters for the proposed analytical model for planar spiral inductors.

PARAMETER	DESCRIPTION	FORMULA
Ar	Area of the inductor	$Ar = s^2$ for square inductors $Ar = \pi \cdot r^2$ for circular inductors
η_{Ar}	Area efficiency : the area occupied by metal compared to the total area	$\eta_{Ar} = \frac{Ar_{metal}}{Ar}$
η_w	Line efficiency : the line width compared to the line pitch (sp is the spacing between two lines)	$\eta_w = \frac{w}{w + sp}$
K_L	Inductor constant	$K_L = 1.3 \cdot 10^{-7}$ H/m
K_R	Resistor constant	$K_R = 3.6 \cdot 10^9 \sqrt{\text{m}} \cdot \text{Hz}/\Omega$
ρ_{metal}	The resistivity of the metal layer	
t_{metal}	The thickness of the metal layer	
t_{ox}	The thickness of the silicon oxide layer between the metal layer and the substrate	
t_{sub}	The thickness of the substrate	
ε_{ox}	The dielectric constant of silicon oxide	$\varepsilon_{ox} = 3.4 \cdot 10^{-11}$ F/m
ω_0	The operating frequency given in [rad/s]	
$R_{\square,sub}$	The sheet resistance of the substrate in [$\Omega/\square$]	

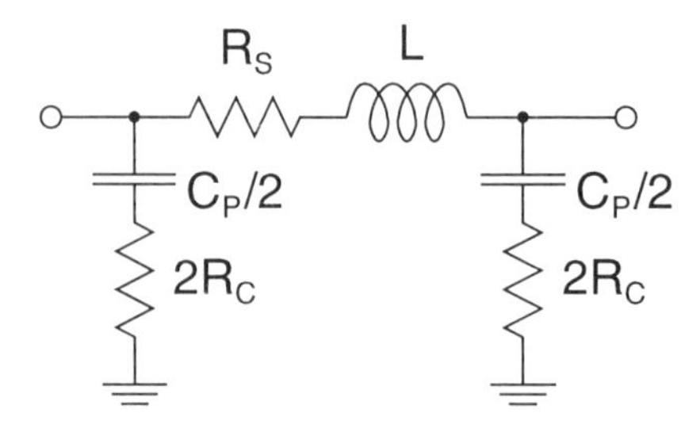

Figure 6.15. The proposed lumped model for spiral inductors.

The series resistance R_S is, for low frequencies, equal to the number of squares times the sheet resistance of the metal conductor [Voorm 1993]. This DC model is extended to incorporate the skin effect which increases the series resistance at high frequencies. The skin effect represents the effect of the magnetic field on the current distribution within a conductor. Normally this magnetic flied is generated by the conductor itself. In a spiral inductor it is the influence of the magnetic field generated by all the conducting turns that has to be taken into account. Here the skin effect is modeled for frequencies up to 3 GHz by means of an equivalent reduced line width δ which depends on frequency and geometry. It models the skin effect with an accuracy of less than 40 %, which is, for resistors, sufficient. The inductance increases also due to the skin effect, but at 3 GHz the increase is still less than 2 %.

The parasitic capacitance C_P is proportional to the occupied metal area. For lowly doped substrates it is important to take into account the resistance in series with this capacitance, R_C, caused by the low conductivity of the substrate. Its actual value depends on the geometry and on the position of grounded substrate contacts. For lowly doped substrates it is approximately 100 Ω (for highly doped substrates it is smaller than 1 Ω). This large value will, at high operating frequencies, reduce the influence of the parasitic capacitor C_P on the design.

6.3.4 Verification of the Model

The presented model for the inductance has been derived by means of both physical interpretation and fittings to the results of the semi-3D electro-magnetic simulation of more than 500 different coils. More information on the used semi-3D electromagnetic simulator can be found in [Freem 1993]. The calculation technique of [Green TPHP74] has been used as extra verification method. Fig. 6.16 shows, as an example, the ratio between the simulated inductance and the inductance predicted by the model versus the inductance value for the simulated planar inductors with both square and circular geometries.

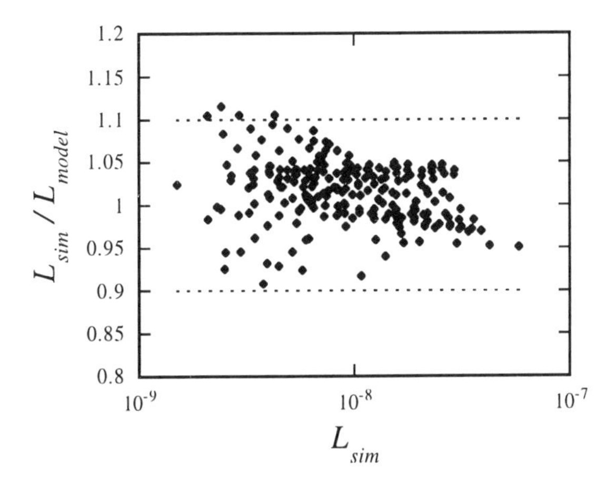

Figure 6.16. Comparison of the inductance model with the results of electro-magnetic simulations (dotted lines indicate 10 % error).

As final verification the model has been compared with the measurement results of several fabricated planar inductors. Table 6.4 gives the comparison between the measured inductance value and the model for three fabricated square inductors. It also gives comparisons between the model and measurement values given in literature [Based ESSC94, Nguy JSSC90, Soyeu EL95]. The calculated values agree well with the measurements, the remaining differences may appear due to differences in layout style, lead inductances and calibration errors (e.g. : in [Nguy JSSC90] the predicted value is given to be 1.3 nH).

The model for series resistance is derived from a DC-model [Voorm 1993] which is extended with the frequency dependent skin effect. The skin effect can, at high frequencies (i.e. 3 GHz), increase the effective resistance to 5 times the DC value. Good modeling of this effect is thus important, but difficult due to its complex dependence on the inductor geometry. A trade-off must therefore be made between a highly accurate and complex or an easy usable model. The skin effect is modeled as an effective line width reduction δ inversely proportional to the operating frequency. δ has again been fitted to the results of the semi-3D electro-magnetic simulation. An accuracy of 40 % for the complete range (up to 3 GHz) is achieved. But, for the most commonly used inductors and for frequencies below 2 GHz the accuracy is better than 20 %. This accuracy is sufficient because the process variation of the sheet resistance is also about 20 %.

Table 6.4. Comparison of the model with measurements and measurement data from literature.

	Ar	w	η_{Ar}	η_w	L_{model}	L_{meas}
coil 1	52900 μm^2	6.4 μm	0.53	0.53	11.3 nH	10.8 nH
coil 2	52900 μm^2	6.4 μm	0.38	0.73	7.01 nH	7.5 nH
coil 3	52900 μm^2	10.0 μm	0.55	0.81	5.37 nH	5.4 nH
[Based ESSC94]	490000 μm^2	90.0 μm	0.85	0.86	4.03 nH	3.7 nH
[Nguy JSSC90]	13200 μm^2	6.5 μm	0.53	0.54	1.38 nH	1.9 nH
[Soyeu EL95]	52900 μm^2	17.0 μm	0.64	0.65	2.35 nH	2.2 nH

6.3.5 Structured Analog Design with Spiral Inductors

The biggest advantage of the presented model is that it gives an analytical expression for the relationship between layout parameters (Ar, w, η_{Ar}, η_w) and electrical parameters (L, R, C). With the proposed model these relationships can be two-way evaluated. Not only do they give a fast result for the electrical properties when the layout is known, they can also be used to directly calculate the layout parameters when the required electrical properties are known. This allows the designer to get a good insight in the design trade-offs and to do his design in a structured way. The proposed model allows for the use of structured analog design techniques to obtain optimal device parameters, but it also allows for the optimization of technologies and the geometry of the planar inductor. In this section examples will be given for these optimizations with the proposed model for planar inductors on the different levels of the design process.

6.3.5.1 Technology optimization. Apart from the obvious statement that the sheet resistance of the conductor used for a planar spiral inductor should be low and that the oxide thickness under the inductor should be high, less obvious technology aspects can be analyzed with the proposed model. One example is whether or not it is better to use only metal levels 2 and 3 (thick oxide, low capacitance) or metal levels 1,2 and 3 (higher capacitance, lower resistance) in a three metal layer process. Fig. 6.17 shows this example. Starting with the assumption that for a certain design an inductor is needed with a given inductance L and a given loss R_S, it is examined which of the

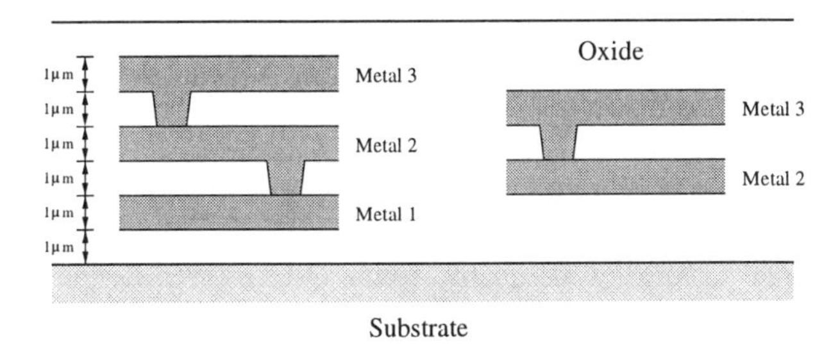

Figure 6.17. Two possible layer configurations for a spiral inductor : three or two parallel metal layers.

two possible situations will render the lowest parasitic capacitance C_P.

$$L_1 = L_2 = L \tag{6.23}$$

$$R_{S,1} = R_{S,2} = R_S \tag{6.24}$$

With the proposed model of section 6.3.3 this can be expanded to (assuming an almost equal area and line efficiency, η_{AR} and η_w for both situations and no skin effect) :

$$\frac{Ar_1^{3/2}}{w_1^2} = \frac{Ar_2^{3/2}}{w_2^2} \tag{6.25}$$

$$\frac{1}{t_{metal,1}} \cdot \frac{Ar_1}{w_1^2} = \frac{1}{t_{metal,2}} \cdot \frac{Ar_2}{w_2^2} \tag{6.26}$$

Substitution results in :

$$\frac{Ar_2^{1/2}}{Ar_1^{1/2}} = \frac{t_{metal,1}}{t_{metal,2}} \quad \Rightarrow \quad \frac{C_{P,2}}{C_{P,1}} = \frac{t_{ox,1}}{t_{ox,2}} \cdot \left(\frac{t_{metal,1}}{t_{metal,2}}\right)^2 \tag{6.27}$$

A general interpretation of this result is that, for the case when the skin effect is small, it is better to make the metal thicker, rather than the oxide. A thicker metal results in a quadratic reduction of the inductor area and therefore also in a quadratic reduction of the parasitic capacitance, while it only scales linearly with the oxide thickness. For the example of fig. 6.17 this means that the first situation, with thicker metal and thinner oxide, gives a 12 % lower parasitic capacitance, although the oxide reduction is much larger (a factor 2) than the increase of the metal thickness (a factor 1.5).

6.3.5.2 Geometry optimization. For inductors there are basically only two electrical parameters which can be chosen (e.g. L and R_S) freely. There are however more layout parameters (circular or square, Ar, w, η_{Ar} and η_w). Geometry optimization reduces this high degree of freedom to two design parameters (Ar and w) by determining the most optimal values for the other parameters. This is done by stating that for a given inductance and a given area, the series resistance should be as low as possible (this is equal to making the quality factor Q at a given frequency as high as possible).

As an example the optimum value for η_{Ar} and η_w is calculated for a square spiral inductor with a given inductance L and area Ar. First it is assumed that there is no skin effect. The model of section 6.3.3 gives the equation for the series resistance R_S.

$$w = K_L^{1/2} \cdot \frac{Ar^{3/4}}{L^{1/2}} \cdot \eta_{Ar}^{5/6} \cdot \eta_w^{1/8} \tag{6.28}$$

$$R_S = \frac{1}{K_L} \cdot \frac{\rho_{metal}}{t_{metal}} \cdot \frac{L}{Ar^{1/2}} \cdot \frac{1}{\eta_{Ar}^{2/3} \cdot \eta_w^{1/4}} \tag{6.29}$$

From this equation it can be concluded that without skin effect the series resistance R_S is minimal when η_{Ar} and η_w are as high as possible, i.e. as close to 1 as possible. In practice this means that the spacing between lines should be minimum size and that the area of the inductor should be fully used. There should be no hole in the middle.

Equation 6.29 does not hold anymore when the skin effect becomes more significant at higher frequencies. The formula for the series resistance is then :

$$R_S = \frac{1}{K_L} \cdot \frac{\rho_{metal}}{t_{metal}} \cdot \frac{L}{Ar^{1/2}} \cdot \frac{1}{\eta_{Ar}^{2/3} \cdot \eta_w^{1/4}} + \frac{\omega_0^2}{K_L^{1/2} \cdot K_R^2} \cdot \frac{t_{metal}}{\rho_{metal}} \cdot L^{1/2} \cdot Ar^{1/4} \cdot \frac{\eta_{Ar}^{1/5}}{\eta_w^{1/8}} \tag{6.30}$$

This equation can be optimized for both η_{Ar} and η_w. It can be seen that η_w should still be as close to 1 as possible. This means that the spacing between the metal lines of a spiral inductor should always be minimal. The optimum value for the hole in the middle of the inductor will depend on the operating frequency ω_0.

$$\eta_{Ar,\mathrm{opt}} = \left(4 \cdot \frac{1}{\omega_0^2} \cdot \frac{K_R^2}{K_L^{1/2}} \cdot \frac{\rho_{metal}^2}{t_{metal}^2} \cdot \frac{L^{1/2}}{Ar^{3/4}} \cdot \frac{1}{\eta_w^{3/8}}\right)^{6/5} \tag{6.31}$$

With η_w almost equal to 1 this means that the spiral inductor should not be filled up to its middle with metal anymore from the following operating frequency on :

$$\omega_0 > 2 \cdot \frac{K_R}{K_L^{1/4}} \cdot \frac{\rho_{metal}}{t_{metal}} \cdot \frac{L^{1/4}}{Ar^{3/8}} \tag{6.32}$$

For instance, for a sheet resistance of 10 m$\Omega/\Box$, an inductance of 5 nH and an area of 1 mm^2, this trade-off point reached at 5.7 GHz operating frequency.

Another interesting point to analyze is whether the square or the near circular spiral inductor layout, as shown both in fig. 6.14, is the most optimal. The same considerations have to be made again : for a given inductance L and a given area Ar, which inductor has the lowest series resistance R_S. From the model equations of section 6.3.3 it can be seen that both are almost equal. Their model only differs in the power of η_w in the expression for the inductance, making the near circular inductor slightly better. However, in the previous paragraph it was already shown that η_w will always have to be very close to 1 and this makes the difference between the two layouts really negligibly small. Nonetheless, in some designs it may be better to compare two inductors with the same inductance L and the same effective occupied area Ar_{eff}. The effective occupied area Ar_{eff} may be larger than the inductor area Ar for a near circular inductor when its empty tips are not used for the layout of any other structure. The difference between the two is :

$$\begin{aligned} Ar &= \pi \cdot r^2 \\ Ar_{eff} &= 4 \cdot r^2 \end{aligned}$$

In this case the square inductor is the preferred one. Its series resistance R_S is 12 % lower than the series resistance of the near circular inductor.

6.3.5.3 Design optimization. During the design and optimization process of an analog integrated circuit a hand-calculation model is very useful. It allows for the development of a forward calculation path which gives an immediate translation of the circuit specifications (such as operating frequency, bandwidth, gain) to the layout parameters (w, l, Ar) of each component. This is a big contrast with the time consuming iterative process of choosing possible layout parameters, running lengthy simulations, analyzing the circuit performance and then choosing new, maybe better layout parameters.

As an example the simplified design of an on-chip buffer which has to drive an off-chip 50 Ω load, is discussed. Fig. 6.18a shows the topology of the output buffer. Fig. 6.18b shows also the parasitic components of the inductor. Half the parasitic capacitor C_P will be short circuited to the power supply. Its resistance R_C is, for this example, assumed to be low. Further on, a parallel resistance R_P is used instead of the series resistance R_S. At a given operating frequency R_P will result in the same loss as R_S when :

$$R_P = \frac{(\omega_0 L)^2}{R_S} \tag{6.33}$$

The buffer will have to realize a specified gain A in a specified bandwidth BW around a specified operating frequency ω_0. The required transconductance g_m of the transistor can already be calculated when it is assumed that the loss in the inductor will be low compared to the load R_L (i.e. $R_L \ll R_P$).

$$g_m = \frac{A}{R_L} \tag{6.34}$$

The bandwidth BW will determine the required inductance L.

$$BW = \frac{R_L}{L} \Rightarrow L = \frac{R_L}{BW} \tag{6.35}$$

The required capacitance on the output node is determined by the operating frequency ω_0.

$$\omega_0 = \frac{1}{\sqrt{L \cdot C}} \Rightarrow C = \frac{1}{\omega_0^2 L} \tag{6.36}$$

This value must be met exactly. It consists of three parts : the load capacitor C_L, which is already fixed, the parasitic capacitances of the transistor on the output node, C_M, and the parasitic capacitance of the inductor, $C_P/2$ ($C = C_L + C_M + C_P/2$). The higher C_P can be, the lower the loss in the inductor will be (assuming no skin effect and a good area and line efficiency) :

$$R_P = \omega_0 \cdot K_L \cdot \frac{t_{metal}}{\rho_{metal}} \cdot \left(\frac{t_{ox}}{\varepsilon_{ox}}\right)^{1/2} \cdot C_P^{1/2} \tag{6.37}$$

The inductor area (and thus its parasitic capacitance C_P) can not be made arbitrarily high. $C_L + C_P/2$ must be lower than C and C_M can only be made low at the cost of a high power consumption. The transistor's g_m is already fixed and a small transistor width w can then only be achieved by using a high $V_{GS} - V_T$. A trade-off has thus to be made here by the designer. The advantage of the model proposed in section 6.3.3 is that it has now on one side a relationship between the chosen inductor area and the signal loss (via equation 6.37) and between the chosen inductor area and the power consumption of the output buffer on the other side. Once the area is chosen the layout parameters of the inductor are fully specified. After this last design step the inductance value and the parasitics of the inductor should always be recalculated with a 3D electro-magnetic finite element simulator in order to obtain a last refinement which will give a more accurate analysis of the specifications which may be expected of the designed circuit.

6.4 CMOS LNA'S

The most critical points for the realization of a highly integrated transceiver in CMOS are the RF input (input of the LNA) and the RF output (output of the upconversion

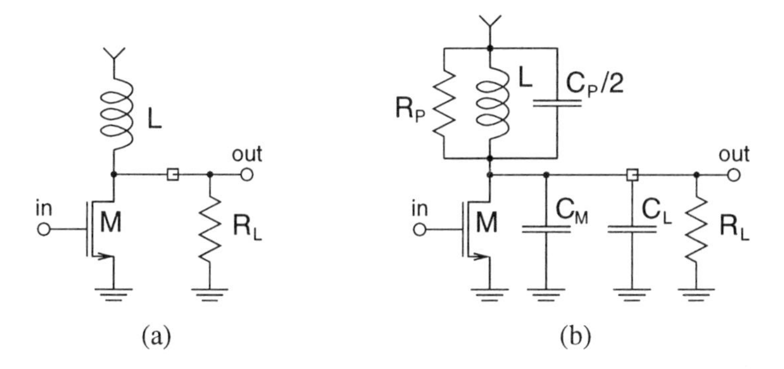

Figure 6.18. The circuit schematic of a) a simplified 50 Ω output buffer with an inductive load and b) the same buffer with its parasitics.

mixers). The main reason for this is that they are external nodes. Internal nodes may have higher impedances which are easier to realize. The realization of output buffers in CMOS was already discussed in section 6.3.1. From this discussion it can be concluded that high frequency silicon bipolar is a more suited technology for the realization of output buffers. It is therefore one of the most difficult building blocks to be realized in CMOS at RF frequencies.

It is often believed that the MOS transistor is also not a good device for the realization of a low noise amplifier (LNA). The reason for this belief is its obviously lower g_m/I_{ds} ratio when compared to a bipolar transistor. A bipolar transistor has a g_m/I value of approximately 40 V^{-1}, while for an nMOS transistor it varies with the $V_{GS} - V_T$ and it is maximally about 7 V^{-1} and this at the cost of having to use a low $V_{GS} - V_T$ which reduces the high frequency capabilities of the MOS transistor. A low noise figure (NF) requires the use of a high g_m and this would lead to the conclusion that a CMOS technology is not well suited for the realization of LNA's at a reasonable power cost. It may however be overhasty to jump to this conclusion.

As MOS transistor gate lengths are sized down to sub-micron and deep sub-micron technologies, the problem of a too low frequency performance at low $V_{GS} - V_T$ values does not hold anymore. An f_{max} of 2 GHz is, especially in deep sub-micron technologies, even at a $V_{GS} - V_T$ of 200 mV available. The realization of an nMOS transistor with a g_m of 1/50 Ω^{-1} (equal to an NF of 2.2 dB [Laker 1994]) would require a current of about 3 mA (assuming a g_m/I of 7 V^{-1} for a 200 mV $V_{GS} - V_T$). This is a value that can be compared to the current consumption of the input stage of reported silicon bipolar LNA realizations [Meyer JSSC74, Rab ACD93]. A similar calculation can however also be made for a bipolar transistor and this gives as result that a current of 375 μA would be required for the realization of a bipolar transistor with a g_m of 1/66 Ω^{-1} (equal to an NF of 2.2 dB [Laker 1994]). From this it could be concluded that a significantly lower power consumption should be expected from bipolar LNA realizations. This is not the case. It is the exponential voltage-to-current law of the bipolar transistor that degrades its capabilities for LNA realizations. LNA's have severe linearity specifications. Its first stage must for instance be able to cope with an input signal of -10 dBm and the gain of the first stage must be high enough (preferably at least 6 dB) in order to reduce the influence of the succeeding stages on the NF. The 1 dB compression point of a bipolar transistor lies however at -18.76 dB (independent of the bias current) [Gilb ACD96], which means that some form of feedback over the first stage must be used to linearize it. This feedback can be realized as emitter degeneration or an overall feedback over the LNA. Both will add noise to the input stage and increase its power requirements. A MOS transistor does not have these problems. It is a much more linear device. Moreover, its distortion components are mainly second order, while the unwanted distortion components in an LNA are mainly of odd order (3rd, 5th, ...). The first stage of a CMOS LNA can therefore be realized as a loaded nMOS without any form of feedback. When the frequency performance of the used technology allows it, this can be done at a very low power consumption. Only an extra inductor, either realized on-chip or as bond wire, for the adaption of the input impedance to 50 Ω has to be added to obtain the full first stage of a CMOS LNA [Sheng ISSCC96, Shaef VLSI96].

There is a second reason why the CMOS transistor is better suited for the realization of LNA's. The equivalent input noise of a bipolar transistor is only in theory and at DC equal to :

$$\mathrm{d}v_{in,eq}^2 = 4kT \cdot \frac{1}{2g_m} \cdot \mathrm{d}f \tag{6.38}$$

In practice the noise of the bipolar transistor will be much higher. First of all there is the base resistance r_b. It has a similar effect as the resistance of the polysilicon gate r_g of a MOS transistor. Difference is that in a MOS transistor r_g is a longitudinal resistor which can be made sufficiently small by using an interdigitated layout with a small value for the width w of each digit. In a bipolar transistor r_b is a vertical resistor which, for a given transistor area, can not be made smaller. Only the use of higher currents and more transistor area, allows for a lower noise due to the base resistance r_b. The excess noise induced by r_b can in a pure bipolar technology still be limited as the value of r_b is often lower than 50 Ω per unit area in these technologies. In BiCMOS technologies the value of r_b can be well over 200 Ω which makes the bipolar transistor in these processes totally unsuited as input transistor for an LNA. An even more important source of noise performance degradation in bipolar transistors is the shot noise induced by the base junction. At low frequencies this can be neglected because of the high value of β_{DC}, but at a high operating frequency f_c the effective value of β is equal to f_T/f_c [Gilb ACD96].

In a MOS transistor all these extra noise sources are not present and the following equation can be taken as a good approximation, even at high operating frequencies [Laker 1994]:

$$dv^2_{in,eq} = 4kT \cdot \frac{2}{3g_m} \cdot df \tag{6.39}$$

Furthermore, with the use of nMOS transistors extra power gain can be obtained by using new circuit topologies that take advantage of current re-use techniques. This is not possible for bipolar devices where feedback techniques are required to overcome its much lower linearity. In an open loop topology transistors can easily be stacked to form one equivalent input transistor which all use the same current. This results in an improved g_m value for the same power consumption. In [Karan ISSCC96] this was done with an nMOS and pMOS on top of each other. Far better results can however be obtained when several nMOS devices are stacked on top of each other. Fig. 6.19 shows three nMOS transistor which are stacked on top of each other together with their resistive loads. Essential for the possible realization of this topology as an input stage for an LNA is the availability of high quality capacitors with low parasitic capacitance to the substrate (lower than 1/10) in the used process. The presented topology for an LNA input stage can, in a sub-micron CMOS topology, be placed within a 3 V power supply and still have a high enough linearity. The equivalent input noise of this input stage is almost 1/3 of the noise of an input stage with the same, but only one, nMOS transistor which would have the power consumption. Signal loss in the capacitors and the bulk effect working on the two top transistors will make the power consumption reduction slightly higher than 1/3, but it is nonetheless again a good example of how nMOS transistors are far better suited for the realization of LNA's than bipolar transistors. The only question which remains is whether or not the nMOS transistors are capable of achieving the high operating frequencies required for wireless application. In sub-micron CMOS technologies and especially in deep sub-micron CMOS technologies this is the case. Today, it is possible to realize in a 0.5 μm CMOS technology with high quality capacitors a 1 GHz LNA with a NF lower than 2.5 dB, an input IP3 of -3 dBm and a gain of 18 dB, using a power supply of 3 V and a power consumption of less than 15 mW.

6.5 QUADRATURE GENERATORS

6.5.1 The RC-CR Allpass Filter

The principle of multi-path topologies for receivers and transmitters was introduced in chapter 3. In multi-path topologies a number of related signals (in general 2 or 4) are put in quadrature and processed in parallel. Multi-path topologies offer many advantages towards full integration and low power consumption. Key issues for the design of multi-path topologies are matching between the different parallel processing paths

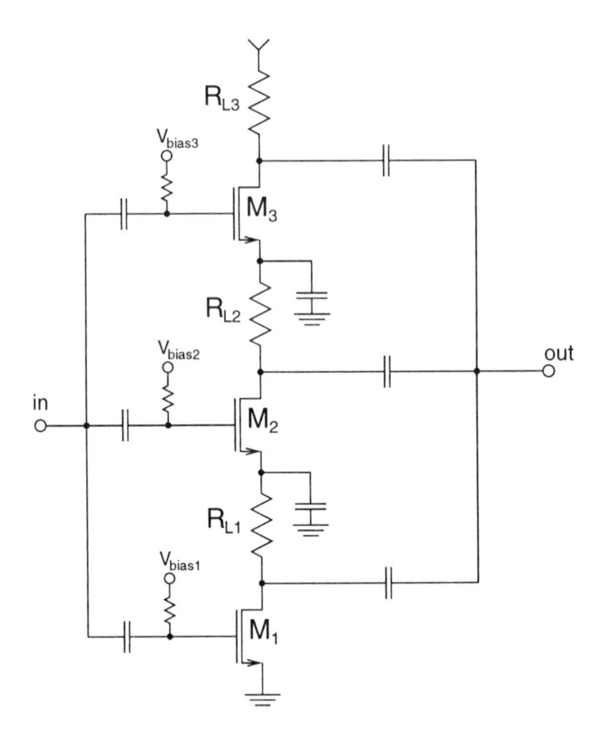

Figure 6.19. The technique of current re-use which reduces the power consumption for the input stage of a CMOS LNA.

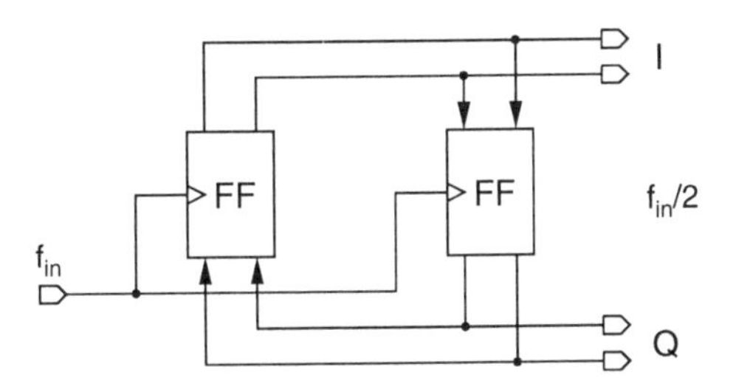

Figure 6.20. The digital generation of quadrature signals.

and the possibility to generate quadrature signals. The absence of good implementation techniques for quadrature generators with which high quality quadrature signals can be obtained, is one of the reasons why multi-path topologies are not widely used today. Only the zero-IF receiver topology, which can accept a limited quadrature accuracy of 25 dB or even less, is an example of a multi-path topology which has been implemented with success [Seven CICC91, Seven ISSCC94, Abidi CICC94, Abidi JSSC95, Abidi ACD96, Abidi ISSCC95].

Today, there are two main techniques known for the implementation of a quadrature generator. One is a digital implementation [Razav ICD96], the other one is the use of an *RC-CR* allpass filter [Stey JSSC92]. Fig. 6.20 shows a possible implementation for a digital quadrature generator. A large advantage of the digital implementation is that a very high quadrature accuracy, even lower than 1° at 1 GHz [Razav ICD96], can be achieved with a highly symmetrical digital circuit topology. The digital implementation however has also many disadvantages. First of all its local oscillator frequency must be twice as high as the operating frequency. A system on 1.8 GHz like for instance DECT will require an oscillator frequency of 3.6 GHz and such a high frequency makes it very hard to implement highly symmetrical digital circuitry. Another disadvantage is that a signal is put in quadrature based on its zero-crossings, meaning that only periodic signals like sines of square waves can be put in quadrature. A pseudo random signal, like the antenna signal, can not be put in quadrature. A third disadvantage is that the signals in quadrature are square wave signals and they are not the most suited signals to be used in receivers and transmitters for mixing because they will also generate mixing at all their odd harmonics.

It is for all these reasons that a continuous time, analog generation of quadrature signals is preferred. An *RC-CR* allpass filter, given in fig. 6.21, is such an analog quadrature generator. It is today the most widely used technique for the implementation of high frequency quadrature generators. At the corner frequency of the RC-pole, $1/2\pi RC$, the input signal will be attenuated with 3 dB and it will have obtained a 45° phase lag. The CR-zero has the same corner frequency, $1/2\pi RC$, and at this frequency the input signal is also attenuated with 3 dB. At this frequency the output signal will have a phase lead of 45°. The two output signals, i.e. the I and Q signal, therefore have at the corner frequency an equal amplitude and a 90° phase difference. Fig. 6.22 shows this transfer function for the RC-pole and the CR-zero. This figure demonstrates that RC-CR structure is not an allpass filter in the strict sense, as known for real signals. It only has a flat amplitude response for its real to complex transfer function. Fig. 6.24 shows these transfer functions, both for positive and negative frequency components. The negative frequency components are in this case unwanted and their amplitude must be compared with the wanted positive frequency component to find the obtained quadrature accuracy.

The problem with the RC-CR topology is that it is very sensitive to mismatch. Relative mismatch between the two resistors R or the two capacitors C will give the pole and zero a different corner frequency, resulting in the fact that there is no frequency anymore at which the I and Q signal have the same amplitude and a 90° phase difference, limiting in this way the lowest level that is reached in the dip of the transfer function for negative frequencies as given in fig. 6.24. This effect is however not the main limiting factor for RC-CR quadrature generators. A 1 % mismatch between the components will still result in a quadrature accuracy of about 40 dB, which is for most applications sufficient.

Loading with parasitic capacitors, caused by the output buffers following the quadrature generator, will in first order not affect the accuracy that can be obtained. It is again only mismatch between the parasitic that will cause a difference between the pole and zero. The only difference is that mismatch between parasitics can be rather large and this is often neglected.

The main problem for the RC-CR quadrature generator is its sensitivity to absolute variations of the R and C values. These variations can be very large from process batch to process batch (up to 20 %) and they result in a difference between the corner frequency and the operating frequency. Fig. 6.24 shows that the bandwidth in which for instance a 25 dB quadrature accuracy can be obtained is very limited. An operating frequency which differs 20 % from the corner frequency, will fall outside this band and the 25 dB accuracy will not be achieved. The consequence is that some kind of tuning or trimming technique will be necessary [Seven ISSCC94]. Another solution can be found in the fact that the quadrature error is mainly an amplitude error. This can be seen from fig. 6.22. Fig. 6.23 shows that the phase difference between the two signals is theoretically equal to 90° from 0 Hz to ∞. The amplitude difference can be corrected by clipping the I and Q signal. In this way quadrature square wave signals are obtained with a good accuracy in a large bandwidth at the cost of having the same disadvantages as with digital quadrature generation.

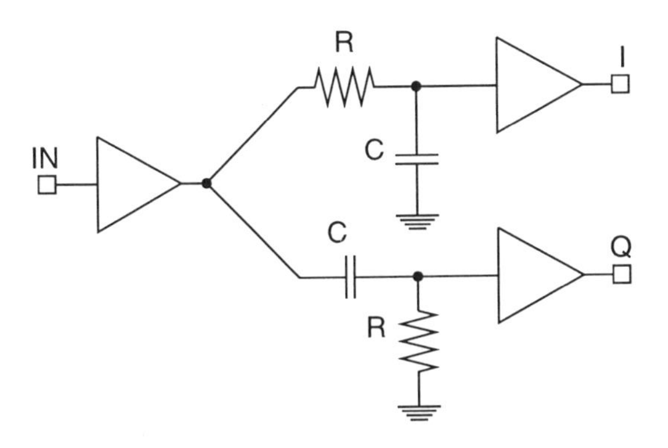

Figure 6.21. The RC-CR allpass filter, classically used as quadrature generator.

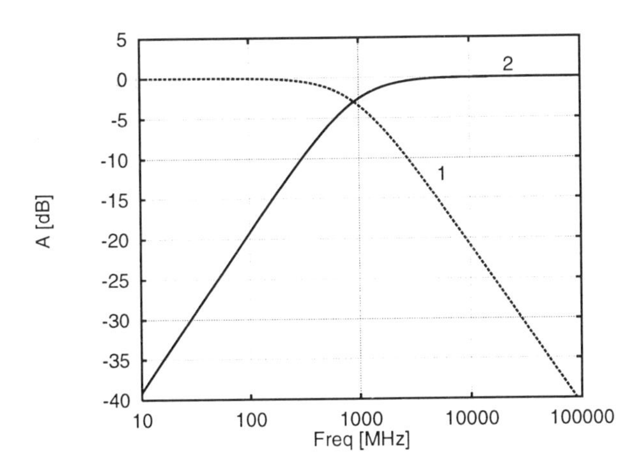

Figure 6.22. The amplitude response of the pole (1) and zero (2) of an RC-CR quadrature generator.

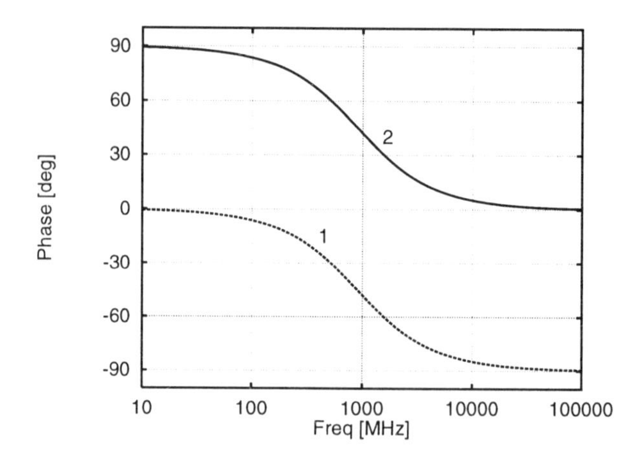

Figure 6.23. The phase response of the pole (1) and zero (2) of an RC-CR quadrature generator.

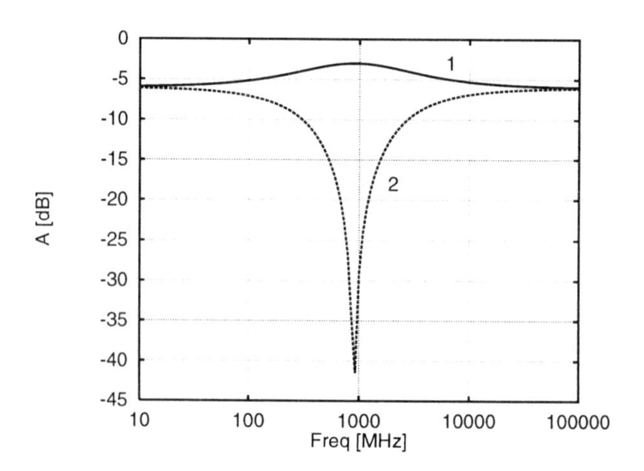

Figure 6.24. Amplitudes of the positive (1) and negative (2) frequency components generated by a perfect RC-CR quadrature generator.

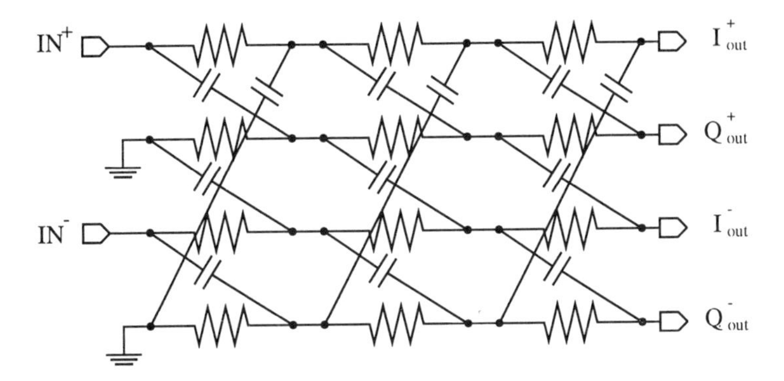

Figure 6.25. A three stage sequence asymmetric polyphase filter used as quadrature generator.

6.5.2 *The Sequence Asymmetric Polyphase Filter*

A sequence asymmetric polyphase filter, described in section 3.3.3.3, can be used as quadrature generator. In that case only one, differential, input signal is applied. The other input is grounded. This is shown is fig. 6.25 for a three stage version.

A polyphase filter has the same properties as an *RC-CR* type quadrature generator. In fact, when only one input is used, while the other three are grounded, it is equal to an *RC-CR* quadrature generator. The difference is that in a polyphase filter several stages can be cascaded. While the phase difference remains theoretically 90° from 0 Hz to ∞, the amplitude error can be made small over a much larger band by cascading several stages with each a different corner frequency. Fig. 6.26 shows this for a three stage polyphase filter with R and C values as given in table 6.5. A quadrature accuracy of 25 or even 30 dB can now be achieved in a large bandwidth around the corner frequency. This makes the polyphase filter insensitive to absolute variations of the R and C values, eliminating in this way the need for any tuning or trimming. Clipping of the output signals is also not necessary anymore which means that not only sinusoidal signals can be put in quadrature. With a polyphase filter any signal, even the antenna signal, can be put in quadrature, opening in this way new opportunities for the realization of new receiver and transmitter topologies. A problem that remains is the sensitivity to mismatch. Fig. 6.27 shows the parasitic transfer function caused by a 5 % mismatch between resistor values for the three stage polyphase filter. The following sections describe how quadrature generation can be made insensitive to mismatch by using new receiver and transmitter topologies.

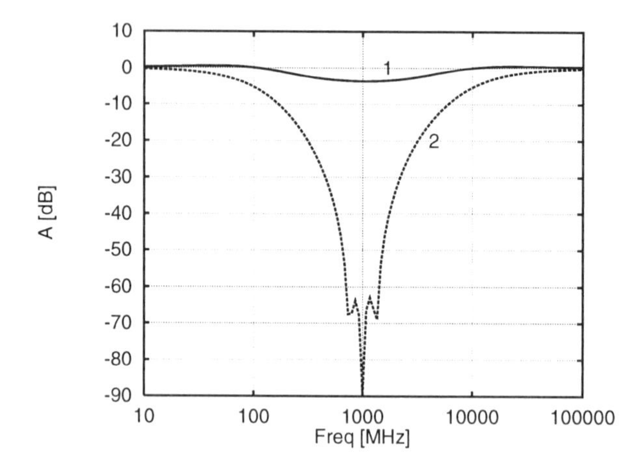

Figure 6.26. The transfer function of a three-stage polyphase filter for positive (1) and negative frequencies (2).

Table 6.5. R and C values for the three stage polyphase filter with center frequency at 1 GHz.

	R	C
Stage 1	120 Ω	1 pF
Stage 2	160 Ω	1 pF
Stage 3	210 Ω	1 pF

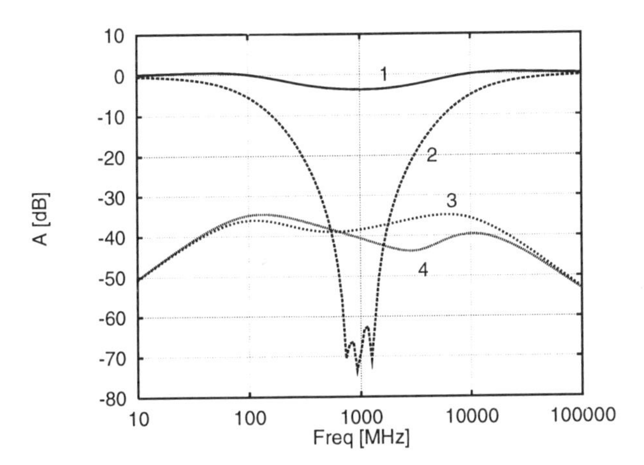

Figure 6.27. The two wanted transfer functions of a three stage polyphase filter (from positive to positive frequencies (1) and from negative to negative frequencies(2)) and the two parasitic transfer functions (from positive to negative frequencies (3) and from negative to positive frequencies (4)) which occur when there is mismatch between its components.

6.5.3 A 900 MHz CMOS Double Quadrature Downconverter

In the last few years the zero-IF downconversion topology is used more and more in receivers for mobile wireless telecommunications applications [Seven CICC91, Abidi CICC94, Min CICC94]. The zero-IF receiver can be implemented with a much higher degree of integration than the conventional IF (heterodyne) receiver. This has made the zero-IF downconverter the preferred topology in the research towards the fully integrated single-chip transceiver.

However, the zero-IF topology has two well-known major drawbacks which makes achieving an acceptable performance with it very hard [Rab ACD93, Balt ACD93]. First, its baseband configuration makes the zero-IF topology highly sensitive to parasitic baseband signals like DC offset voltages and self-mixing products caused by crosstalk between the RF and the LO signals. The second source of performance reduction is inherent to any analog integrated multi-path topology. Excellent matching between the different downconversion paths is required, but limited in analog implementations. The effects of mismatch, i.e. phase and amplitude errors, degrade the signal quality because they result in a reduced mirror signal suppression.

The zero-IF receiver topology is an example of a basic multi-path topology. In chapter 3 it was demonstrated that several other multi-path topologies are possible, opening in this way many possibilities for the realization of receiver topologies which combine the property of a very good integratability with the ability to achieve a very good performance. Where in the zero-IF receiver the performance is limited due to the limited matching accuracy of the *RC-CR* quadrature generator, here the chip realization of a new multi-path type receiver topology is proposed which is highly insensitive to mismatch and process variations and which at the same time eliminates the problem of the sensitivity to parasitic baseband signals.

The low-IF receiver, presented in section 3.5.1.2, is, like the zero-IF receiver, also a multi-path topology which can be implemented in a highly integrated way. It uses an IF of a few hundred kHz and is therefore not sensitive to parasitic baseband signals like DC offset voltages and self-mixing products. In this way the low-IF receiver combines the advantages of both the IF and the zero-IF receiver. It can have a high performance and a high degree of integration at the same time. The main drawback of the low-IF receivers is that they are more sensitive to a good mirror signal suppression, because, in contrast to the zero-IF receiver, here the mirror signal can be larger than the wanted signal. In the previous section it was demonstrated that the phase error of the quadrature generator is the main limitation for the mirror signal suppression in the classical downconverter used in zero-IF receivers. A typical quadrature downconverter with a phase error of 3° [Stey JSSC92], results in a maximal mirror signal suppression of 26 dB. Therefore, a new fully integrated CMOS quadrature downconverter with a phase accuracy of 0.3^o has been developed. It is a key building block for the realization of high quality low-IF receivers. The proposed quadrature downconverter can form in combination with an LNA and a synthesizer the complete and fully integrated analog low-IF receiver front-end.

6.5.3.1 The Double Quadrature Structure. Fig. 6.28a shows the conventional quadrature downconverter. The LO signal is put into quadrature and the downconversion is done twice. Signals situated at a frequency higher than the LO will be downconverted to positive low frequencies. Signals lower than the LO will be downconverted to negative frequencies. With a phase error on the LO signal, signals coming from one side of the LO will also have a signal component coming from the other side of the LO. This is mirror signal crosstalk. This parasitic crosstalk is, according to equation 3.28, equal to $\tan(\Delta\phi)$. For an *RC-CR* quadrature generator this phase error is determined by the mismatch between the cut-off frequencies f_c of the pole and the zero of the *RC-CR* structure. For a certain mismatch Δf_c, $\tan(\Delta\phi)$ is maximal at f_c and equal to :

$$\tan(\Delta\phi) = \frac{\Delta f_c}{2 f_c} \tag{6.40}$$

So, a phase error of less than 0.3^o requires that the matching between the RC and CR network is better than 1 %. This value is then still only achieved in a very small band around f_c. Further away from f_c the amplitude mismatch of the *RC-CR* structure will dominate, resulting in a very high sensitivity to absolute parameter variations. The

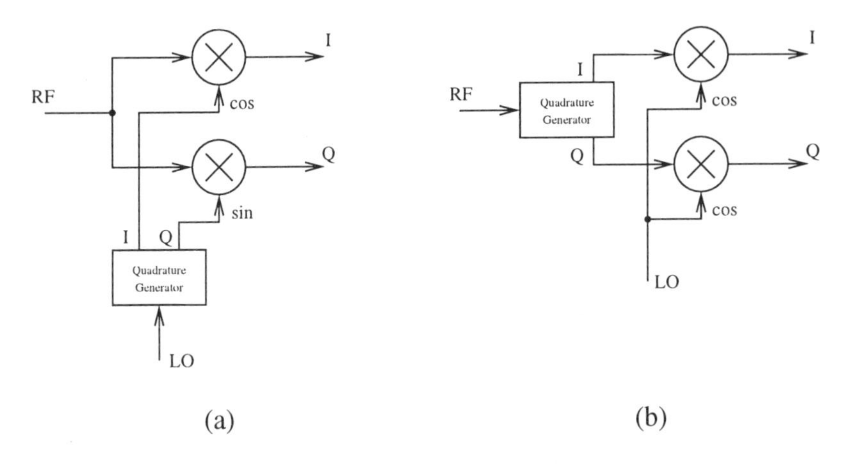

Figure 6.28. a) The conventional quadrature downconverter and b) an alternative structure.

limited amplitude matching is often corrected by clipping the quadrature LO signals. A good phase accuracy (better than 1°) can with the *RC-CR* structure only be achieved by means of extensive tuning and trimming [Seven ISSCC94].

Fig. 6.28b shows an alternative topology for the conventional quadrature downconverter [Taka ISSCC95]. The RF signal is now put into quadrature and both the HF I- and Q-signals are downconverted with a sine. The effect of a phase error in the quadrature generator is the same : a mirror signal crosstalk equal to $\tan(\Delta\phi)$. The difference is that this structure can not be used with an *RC-CR* quadrature generator, as clipping of the RF signal is highly unwanted.

The topology in fig. 6.29 gives the newly presented quadrature downconverter, it is called the double quadrature downconversion structure. It is a combination of the structures of fig. 6.28 : both the RF and the LO signal are put into quadrature. The crosscoupling and combination of the four mixing products makes that the amplitude and phase errors are extra suppressed. The mirror signal crosstalk due to phase and amplitude errors for this structure is :

$$\tan(\Delta\phi) = \tan(\Delta\phi_{RF}) \cdot \tan(\Delta\phi_{LO}) + \frac{\tan(\Delta\phi_{Mixers})}{2} \tag{6.41}$$

$$\frac{\Delta A}{A} = \frac{\Delta A_{RF}}{A_{RF}} \cdot \frac{\Delta A_{LO}}{A_{LO}} + \frac{\Delta G}{2G} \tag{6.42}$$

G is the conversion gain of the downconversion mixers.

Hence, the double quadrature structure is highly insensitive to mismatches in the quadrature generators. Its mirror signal suppression is therefore determined by the

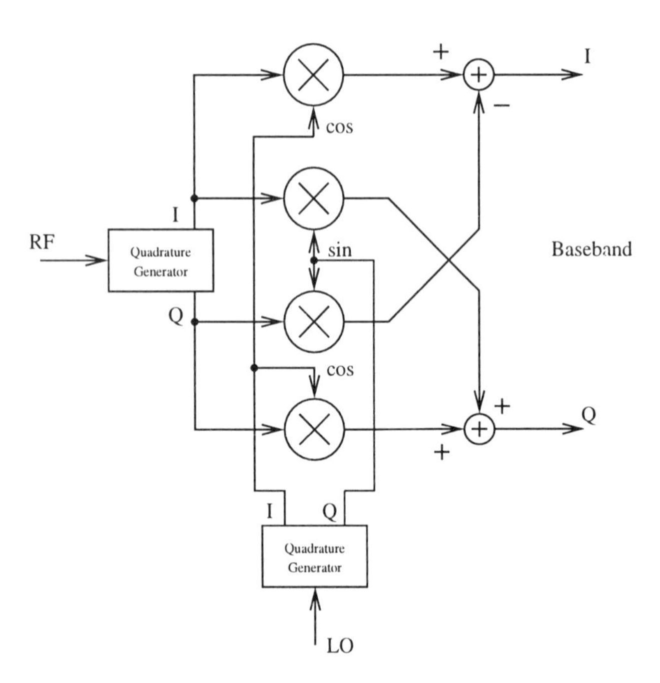

Figure 6.29. The double quadrature downconverter topology.

matching between the downconversion mixers. The phase mismatch between the downconversion mixers is approximately (f_{BW} is the input bandwidth of the mixers) :

$$\tan(\Delta\phi) \approx \frac{f_{LO}}{f_{BW}} \cdot \frac{\Delta f_{BW}}{f_{BW}} \tag{6.43}$$

This means that the phase error can be made low when the mixers have a large input bandwidth. The amplitude error between the mixers is determined by the mismatch in DC conversion gain. It gives a mirror signal crosstalk equal to $\Delta G/2G$ and it is therefore the main remaining source of mirror signal crosstalk. A good design with large enough mixing transistors will significantly reduce this effect and any remaining amplitude error can easily be eliminated by performing a closely matched digital AGC operation on the sampled low frequency I- and Q-signals in the DSP which does the further processing and demodulation of these signals.

6.5.3.2 The Broadband Quadrature Generator. The double quadrature structure requires the use of two quadrature generators which still have to have a relatively good amplitude and phase matching in a broad passband. Making the quadrature generators broadband, eliminates the need for any frequency tuning to compensate for absolute parameter variations. Sequence asymmetric polyphase filters, as presented in section 3.3.3.3, can be used for this purpose.

One stage of a passive sequence asymmetric polyphase filter, as shown in fig. 6.30, passes positive sequences and suppresses negative sequences at $1/2\pi RC$. The big advantage of this structure over the classical *RC*-*CR* quadrature generator is that several stages can be cascaded. This allows for the synthesis of wideband filters, which largely reduces the the sensitivity to absolute parameter variations. Fig. 6.31 gives the simulation results for the positive and negative frequency response of a two stage version. The two stages have different corner frequencies and each dip in the transfer function for negative frequencies lies at the corner frequency, $1/2\pi RC$, of one of the stages.

6.5.3.3 Full CMOS Realization. Fig. 6.33 shows the block schematic of the realized chip. It has been realized in a standard 0.7 μm CMOS technology and does not require any external components, nor does it require any tuning or trimming. First, a single-ended to differential conversion is performed on the RF and LO signal. The circuit schematic of this single-ended to differential converter is given in fig. 6.34. The first stage, with transistor $M1$, performs the conversion. The low-ohmic poly resistors ensure the linearity of the conversion. The dummy transistor $M2$ balances the positive and negative output node. This limits the phase error of the 180° to about 3° at 1 GHz. Transistors $M3p$ and $M3n$ form resistor loaded amplifiers. They compensate for the signal loss of the single-ended to differential stage. The resistor loaded single

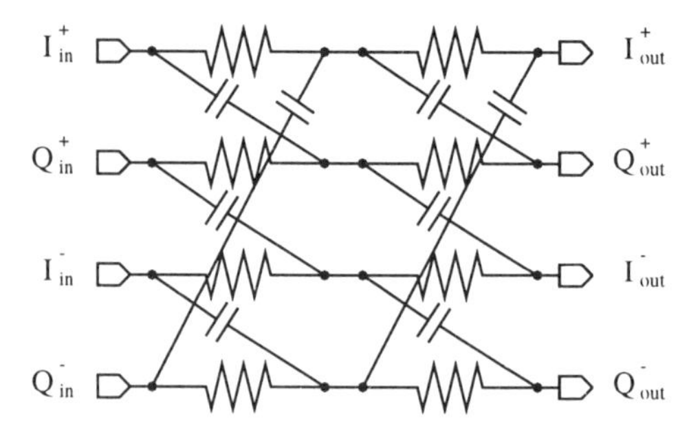

Figure 6.30. The two-stage passive sequence asymmetric polyphase filter, used as quadrature generator.

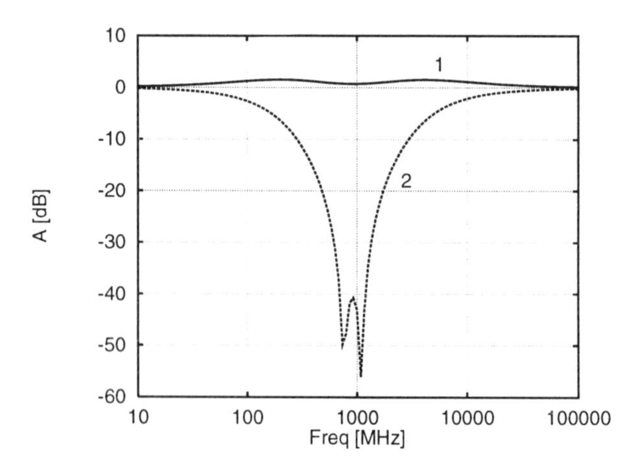

Figure 6.31. The transfer function for positive and negative frequencies of a two-stage passive sequence asymmetric polyphase filter.

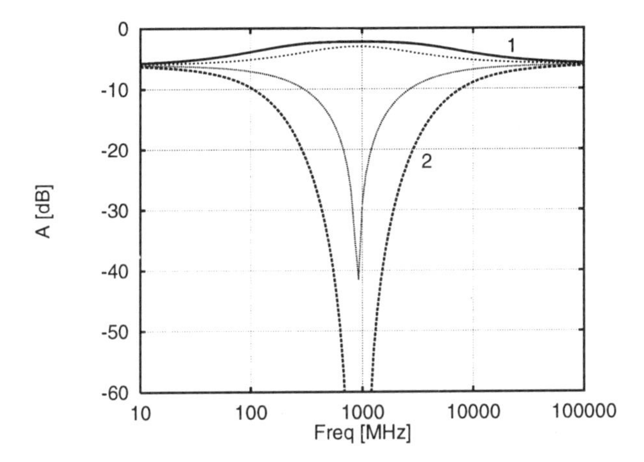

Figure 6.32. The transfer function for the wanted signal (1) and the mirror signal (2) for the complete double quadrature downconverter. The thin lines give, for comparison, the wanted and mirror signal transfer function for a quadrature downconverter with an ideal RC-CR quadrature generator.

transistor amplifier structure is the only one which can still process 1 GHz signals in 0.7 μm CMOS. It does not have a good linearity, but its main distortion components are second order and the second harmonic distortion components are, in this topology, common-mode signals. The common-mode signals pass through the succeeding polyphase filter and are then suppressed at the input of the double balanced downconversion mixers.

Two two-stage sequence asymmetric polyphase filters are implemented on the chip. Their passband is designed to range from about 600 MHz to 1.2 GHz. The signals at 900 MHz are in this way always set correctly in quadrature, independent of process variations. Amplifier stages are again added after the polyphase filters to compensate for the 6 dB signal loss introduced by the polyphase filter.

The mixers are realized with the highly linear CMOS topology for downconverters presented in section 6.2.2. The topology presented there has been redesigned for the 0.7 μm CMOS process. A high input bandwidth can be achieved with this topology by using small modulating transistors (by applying a large $V_{GS} - V_T$) and large capacitors on the virtual ground nodes. The large capacitors make that the opamp only has to process low frequency signals. This makes it possible to use an opamp with large input transistors, which results in a low noise despite its low power consumption. The four fully balanced downconversion mixers are realized with sixteen HF modulating transistors and two LF opamps.

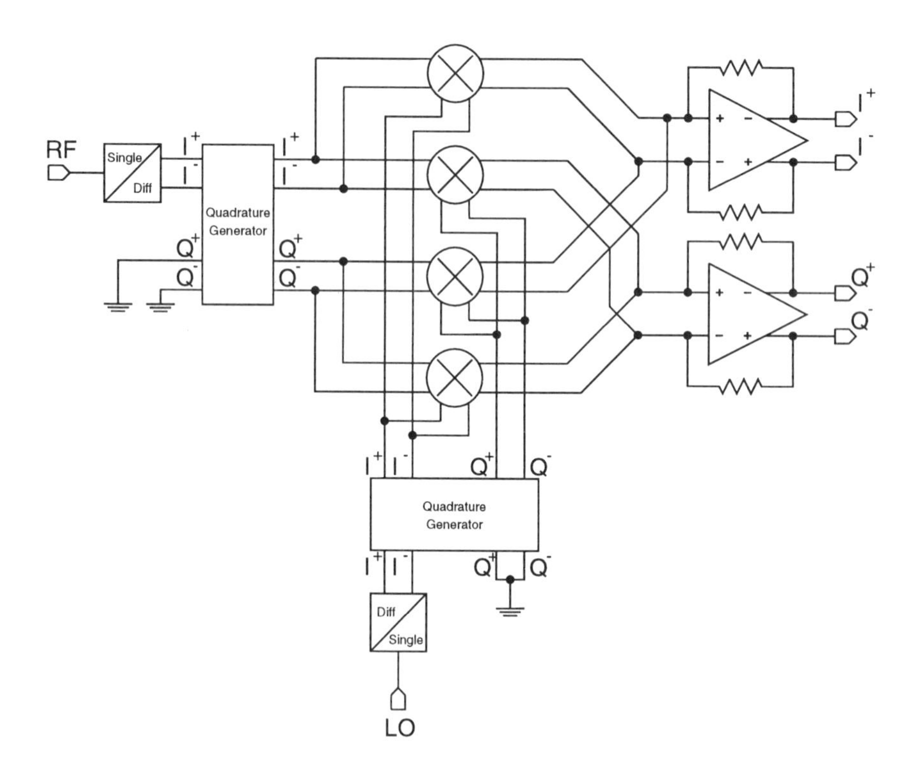

Figure 6.33. Block schematic of the realized CMOS receiver front-end chip.

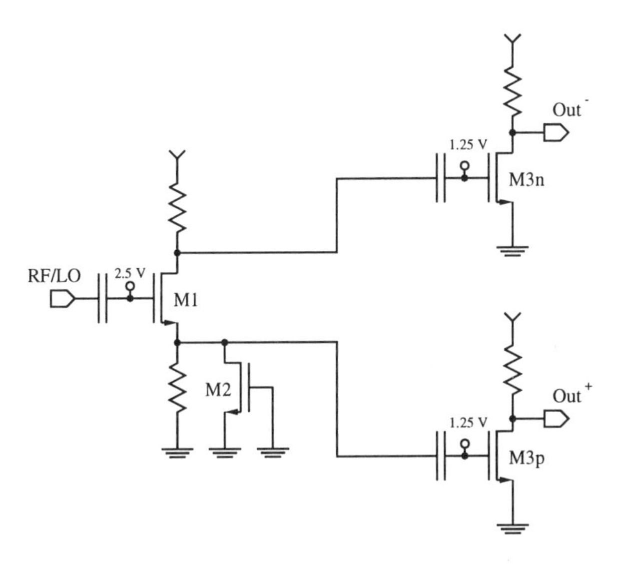

Figure 6.34. The singe-ended to differential converter.

6.5.3.4 Measurement Results of the CMOS Front-End Chip. Fig. 6.35 shows a microphotograph of the realized chip [Crols ISSCC95]. It is realized in a standard 0.7 μm CMOS process and measures 6 mm^2. It operates on a single 5 V power supply voltage and consumes 500 mW, mainly dissipated in the HF resistor loaded amplifiers. The measured conversion gain decays from 16.4 dB at low frequencies to 9.2 dB at 900 MHz. This is equivalent to a 3.1 dB HF signal loss for both the RF and LO signal at 900 MHz. The measured output bandwidth is 3.09 MHz. Fig. 6.36 shows the third-order interception point (IP3) measurement. The IP3 is 27.9 dBm, the measured noise figure (for a 12 dBm LO signal) is 24.0 dB. The measurements of the phase and amplitude errors in function of frequency are given in fig. 6.37. The quadrature output signal is a low frequency signal and its phase and amplitude erros can therefore be measured with a very high accuracy by menas of a low frequency two-channel dynamic signal analyzer. The phase error is less than 0.3° in the 500 to 900 MHz band, which is as expected. In this band the amplitude error is situated at 0.5 dB. These phase and amplitude errors result, after a digital AGC operation, in a mirror signal suppression of 46 dB. Both phase and amplitude errors totally depend on matching. This measurement has therefore been performed on several samples and all samples show the same or better phase and amplitude errors.

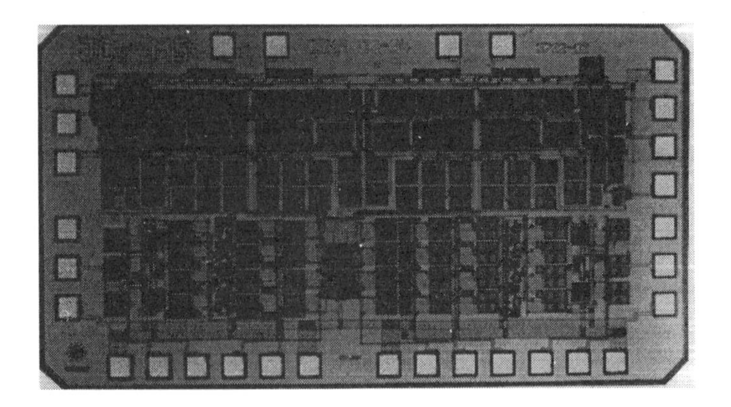

Figure 6.35. Microphotograph of the CMOS double quadrature downconverter chip.

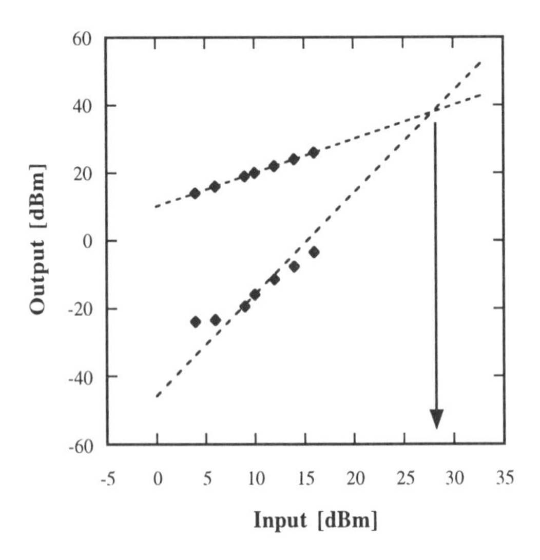

Figure 6.36. The third-order interception point measurement (IP3).

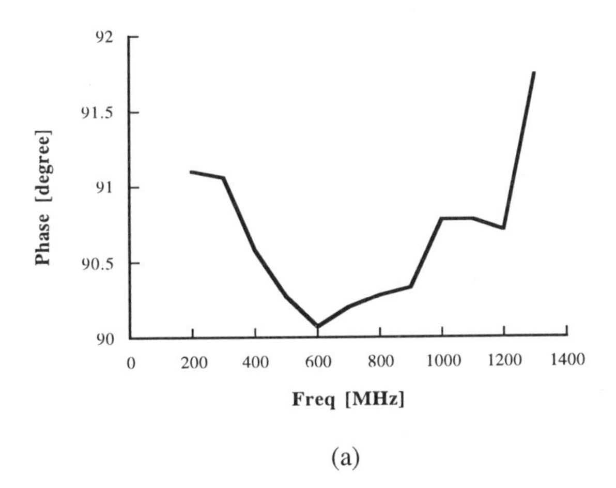

(a)

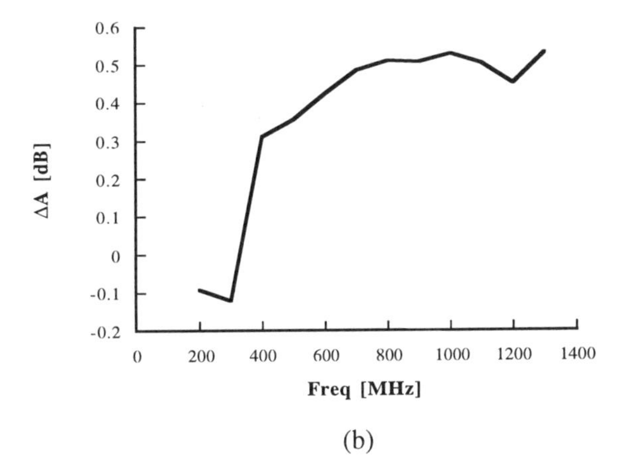

(b)

Figure 6.37. a) The measured phase error and b) the measured amplitude error versus frequency.

6.5.4 A 900 MHz CMOS Double Quadrature Upconverter

6.5.4.1 The Double Quadrature Topology for Upconversion. Today, direct upconversion, shown in fig. 6.38, is used more and more for the realization of transmitters. Direct upconversion, with a quadrature mixer, has the advantage that it can be integrated in a much better way than the heterodyne upconverter which needs intermediate frequencies and off-chip bandpass filters to suppress mirror signals generated during upconversion. In a direct quadrature upconverter the baseband I and Q signal are multiplied with respectively the local oscillator I and Q signal. This directly upconverts the baseband signals to their wanted carrier frequency, which is equal to the frequency of the local oscillator signals. There is no need anymore for the suppression of mirror signals. In the case of direct upconversion the upper and lower sideband are each other's mirror and sideband interference is prevented by doing the upconversion in quadrature. The sideband separation that can be achieved depends on the accuracy of the quadrature generator for the local oscillator signal and, as discussed in sections 6.5.1 and 6.5.3.2, this can be rather limited for the classically used *RC-CR* type quadrature generator.

The double quadrature double quadrature upconversion structure, previously introduced for downconversion, can also be used in direct upconversion to obtain a high quadrature accuracy in a large bandwidth, independent of mismatch. Fig. 6.39 presents the double quadrature topology for upconversion. Each baseband signal is multiplied with each local oscillator signal. The four outputs are summed two by two, delivering a high frequency quadrature signal which has, ideally, no negative frequency components (meaning that the obtained I and Q signal are perfect 90° phase shifted versions of each other). Quadrature inaccuracies of the local oscillator signal and mismatch between the I and Q paths will reduce the quality of both the obtained I and Q signals, but in the presented double quadrature topology, a polyphase filter is used in the high frequency output signal path to further filter these remaining inaccuracies. The consequence is that the quadrature specifications on both the quadrature generator for the local oscillator signal and for the high frequency polyphase filter can be relaxed while still a wide bandwidth can be achieved. The big difference between double quadrature upconversion and double quadrature downconversion is that the polyphase filter is used here as quadrature filter (with 2 inputs and 2 outputs) and not as quadrature generator (with 1 input and 2 outputs).

The remaining quadrature error of the double quadrature upconversion topology is mainly determined by the mismatch between the four upconversion mixers, resulting in phase accuracies far better than the 3° of a classical *RC-CR* type of quadrature generator. Equations 6.44 and 6.45 give an expression for respectively the phase and amplitude error for the double quadrature upconverter topology in transmitters.

$$\tan(\Delta\phi) = \tan(\Delta\phi_{RF}) \cdot \tan(\Delta\phi_{LO}) + \frac{\tan(\Delta\phi_{Mixers})}{2} \qquad (6.44)$$

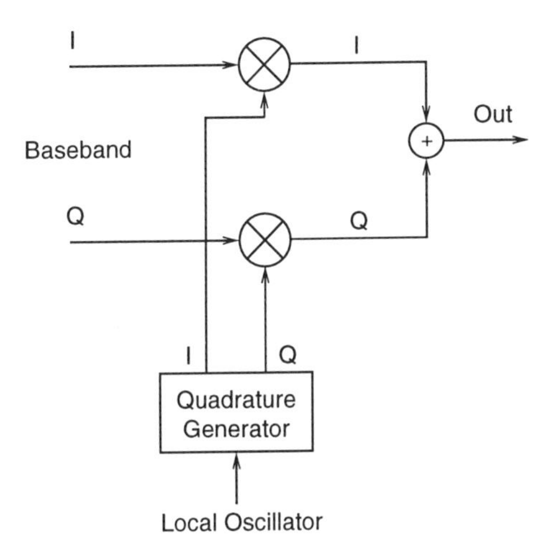

Figure 6.38. The quadrature topology for direct upconversion.

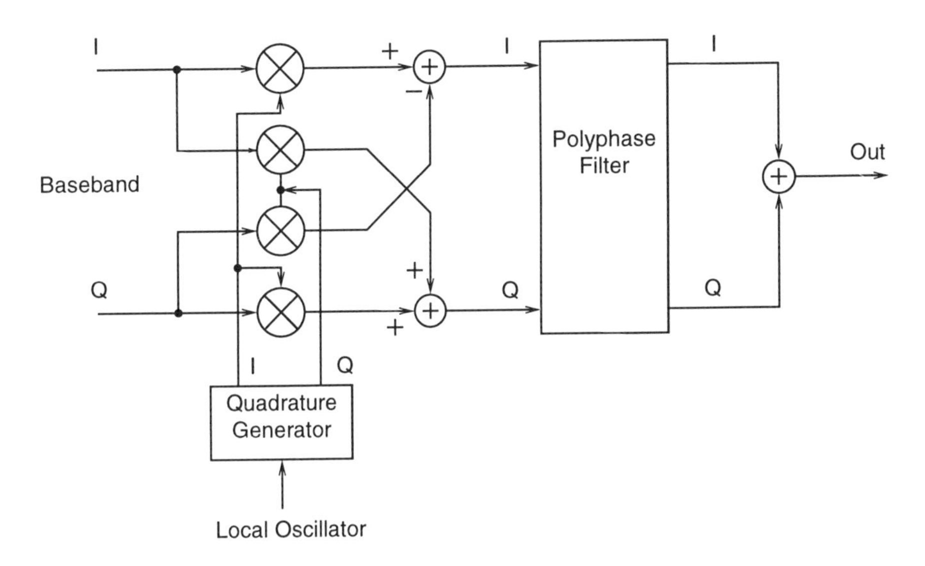

Figure 6.39. The double quadrature topology for upconversion.

$$\frac{\Delta A}{A} = \frac{\Delta A_{RF}}{A_{RF}} \cdot \frac{\Delta A_{LO}}{A_{LO}} + \frac{\Delta G}{2G} \tag{6.45}$$

These equations are the same as for the double quadrature downconversion system (see equations 6.41 and 6.42). The phase and amplitude errors of the LO quadrature generator and the polyphase filter are multiplied, meaning that the polyphase filter further reduces the errors of the quadrature generator, even when it has itself only a limited quadrature accuracy. The matching accuracy of the mixers becomes, as a result, dominant, but this can be made sufficiently low by doing a design towards minimization of transistor mismatch [Pelgr JSSC89, Bast ICMTS95].

6.5.4.2 Circuit Description. The upconverter mixers, shown in fig. 6.40, are true analog multipliers. Four of these mixers are implemented on chip. They perform a multiplication of the differential baseband signals (BB) with a single-ended sinusoidal local oscillator signal (LO) and they implement at the same time the single-ended to differential conversion for the high frequency signal. Both the low frequency and the high frequency input are highly linear. This is obtained by using a MOS mixing transistor in the linear region. Via on-chip inductors the baseband signal is differentially applied between the source and drain of the mixing transistor. The applied voltage across the transistor is converted into a current, dependent on the gate voltage which is modulated with the single-ended local oscillator signal. The differential structure for the baseband signals enables the use of single-ended local oscillator signals and eliminates the problem of the dependence of the transistors g_{DS} on its V_{DS}. The high frequency output currents are buffered by means of cascode transistors through which they are provided to the output polyphase filter.

The use of a true multiplier for the upconversion has, contrary to commutating mixers, the advantage that it produces far lower harmonic components at three and five times the carrier frequency. Such components have to be filtered out with an external ceramic filter before the signal is sent to the power amplifier. With true multiplication this filter can be omitted. The disadvantage of true multiplication is that the efficiency is slightly (1.4 times) lower than the efficiency of a commutating mixer (see section 6.2.2.4). The presented topology of fig. 6.40 is basically a passive structure. None of the transistors implements an amplification function and the conversion gain of the proposed upconversion mixer will therefore be low.

An upconverter test-chip with the proposed architecture has been implemented in a 0.7 μm CMOS technology (the process parameters are described in appendix A). In table 6.6 the dimensions of the realized devices are given. The use of the 0.7 μm CMOS process has some drawbacks. First of all its capacitors have poor quality. They are of the poly-to-diffusion type and their parasitics to the substrate can be as high as 25 % of the capacitor value. This is mainly a problem for the polyphase filter which uses these capacitors. This filter has to be driven from the output stages of the mixers and it reduces in this way significantly the frequency behavior of the mixer output

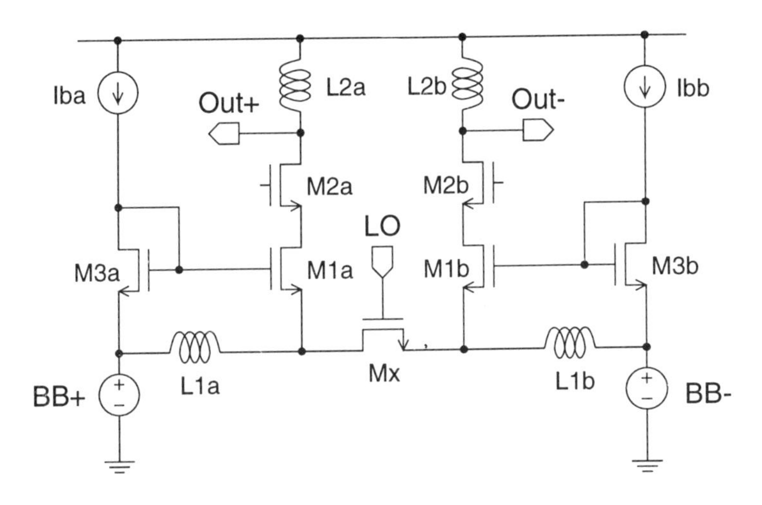

Figure 6.40. Circuit diagram of the upconversion mixer.

stages. To compensate these capacitors on-chip inductors have been used. The 0.7 μm process has however a highly doped substrate which results, due to the generation of eddy-currents, in an increase of both the resistive and capacitive parasitics of the inductor. It is therefore not possible to realize a high efficiency upconversion mixer in the used 0.7 μm CMOS process. Nonetheless, the performance of the double quadrature structure is independent of the actual signal amplitudes and with the proposed mixer topology, which gives a flat transfer function around 900 MHz, the double quadrature structure for upconversion can be demonstrated, however with a limited conversion gain of about -25 dB. With the further evolution towards deep-submicron technologies, the realization of CMOS SSB upconverters with high efficiency will become feasible [Stey ACD96].

The output currents of the upconversion mixers are summed two by two and fed into the two-stage polyphase filter. After the polyphase filter the output currents are again combined into a single balun,on-chip single-ended signal. This is done by means of an on-chip balun, shown in fig. 6.41, which drives a 50 Ω load. A two-stage filter is used in order to obtain the high quadrature accuracy in a wide band, making the circuit insensitive to absolute variations of capacitors and resistors, resulting in no need for tuning or trimming. For the quadrature generation of the local oscillator signal the classical, first-order, *RC-CR* circuit is used. The *RC-CR* quadrature generator has only a limited accuracy, but this is more than sufficient in the double quadrature topology. In combination with the two-stage output polyphase filter this results in a transfer function for quadrature signals which is equivalent with three stages, however, thanks to the double quadrature structure, without the normal sensitivity to capacitor and

Table 6.6. Device sizes for the double quadrature upconverter.

DEVICE	TYPE	SIZES
M1	nMOS	$w = 1200\ \mu m$, $l = 0.7\ \mu m$
M2	nMOS	$w = 135\ \mu m$, $l = 1.5\ \mu m$
M3	nMOS	$w = 120\ \mu m$, $l = 0.7\ \mu m$
Ib	Current source	I = 2 mA
L1	Inductor	$Ar = 342000\ \mu m$, $w = 64.0\ \mu m$, $\eta_{Ar} = 0.74$, $\eta_w = 0.98$
L2	Inductor	$Ar = 347000\ \mu m$, $w = 70.0\ \mu m$, $\eta_{Ar} = 0.81$, $\eta_w = 0.98$
K1	Balun	$Ar = 168000\ \mu m$, $w = 18.0\ \mu m$, $\eta_{Ar} = 0.89$, $\eta_w = 0.92$
R1	Resistor	$R = 140\ \Omega$
R2	Resistor	$R = 102\ \Omega$
R3	Resistor	$R = 68\ \Omega$
C1	Capacitor	$C = 1.25$ pF
C2	Capacitor	$C = 2.2$ pF
C3	Capacitor	$C = 2.2$ pF

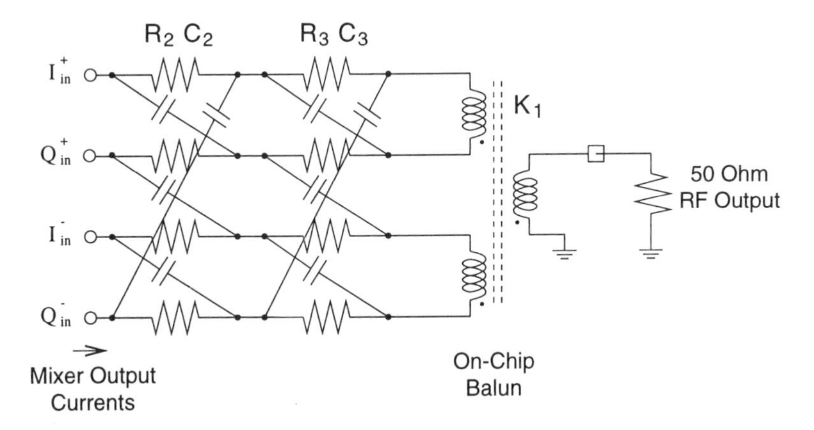

Figure 6.41. Implementation of the two-stage polyphase filter and the on-chip balun which brings the output signal of chip.

resistor mismatch. The advantage of the *RC-CR* structure is that it can be used to generate quadrature signals from a single-ended signal.

6.5.4.3 Measurement Results. Fig. 6.42 shows the chip-microphotograph of the implemented double quadrature upconverter [Crols ESSCC96]. It is realized in a 0.7 μm CMOS process. The die size is 16 mm^2, mainly occupied by the 12 on-chip inductors and the on-chip balun placed at the output. Fig. 6.43 shows the measured sideband suppression in function of frequency. It shows a sideband suppression of more than 30 dB over a 200 MHz band from 700 to 925 MHz. This corresponds to a phase error of less than 1° and an amplitude error of less than 10 mdB.

6.6 LOW FREQUENCY ACTIVE INTEGRATED POLYPHASE FILTERS

6.6.1 Active Integrated Polyphase Filters

The active integrated sequence asymmetric polyphase filter has been introduced in section 3.3.3.2. It is an essential part for the realization of an analog low-IF receiver. The topology of such a low-IF receiver is given in fig. 6.44. It is due to the selectivity of the sequence asymmetric polyphase filter between positive and negative that the mirror signal suppression can be postponed from high frequency to the low frequency signal processing part after the downconversion. It allows in this way for the use of a low IF frequency (a few hundred kHz), which in turn results in a very good integratability of both the HF and LF part of the receiver without having any problems related to baseband operation. An active integrated polyphase filter can however not only perform the mirror signal suppression. It can also perform the channel selection

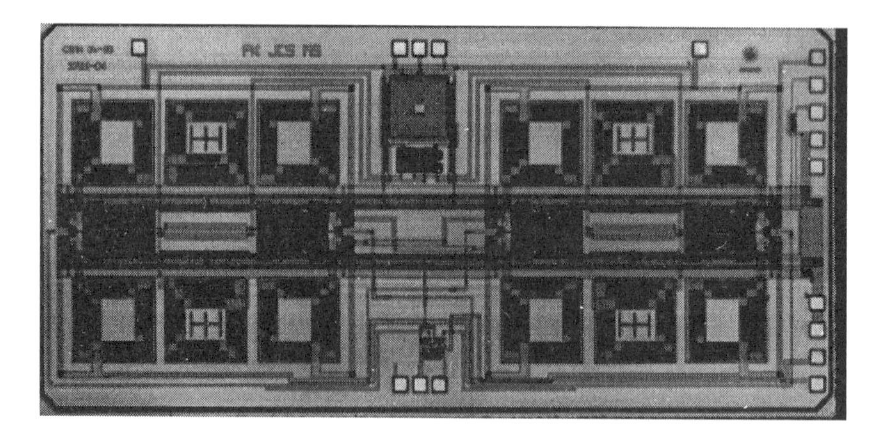

Figure 6.42. Microphotograph of the double quadrature upconversion chip.

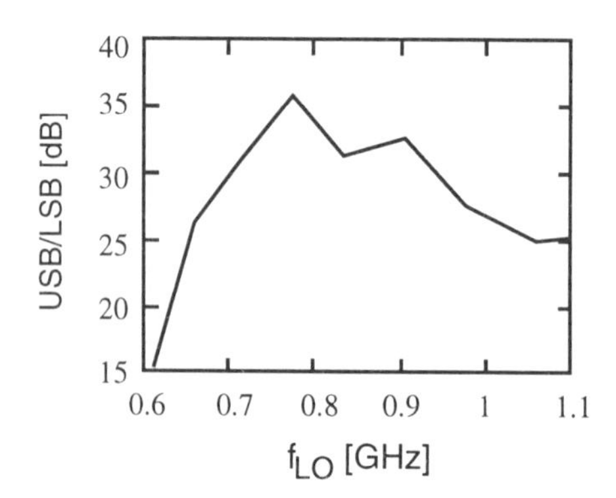

Figure 6.43. The measured sideband suppression in function of frequency.

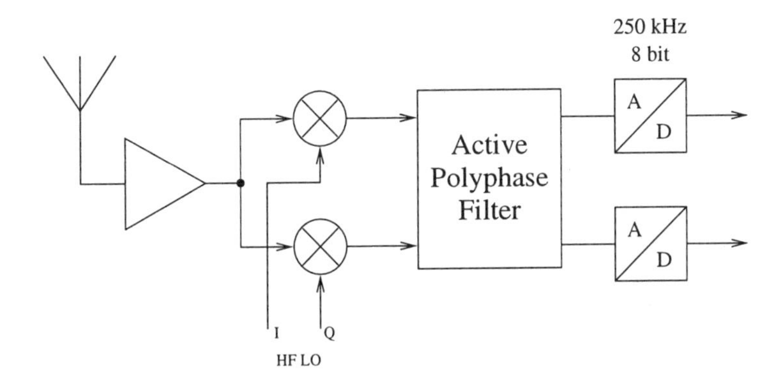

Figure 6.44. The active integrated low frequency polyphase filter used in a low-IF receiver.

and the suppression of adjacent channels which is involved in this process. This is different from the passive polyphase filters (described in section 3.3.3.3).

The most important application for polyphase filters is the suppression of the negative or positive frequency component of a complex signal. This can be done with a bandpass filter which results from the linear transformation of a lowpass filter. The classic lowpass to bandpass transformation does not change the real properties of the lowpass filter and it gives the bandpass filter a passband around $\omega = +\omega_c$ and $\omega = -\omega_c$. After a linear frequency transformation, given in equation 6.46 the bandpass filter has the lowpass filter characteristic only around $\omega = +\omega_c$.

$$\mathrm{j}\omega \rightarrow \mathrm{j}\omega - \mathrm{j}\omega_c \tag{6.46}$$

This transformation can only be realized as a polyphase filter because it introduces complex coefficients in the rational polynomial transfer function of the filter. Equation 6.47 gives this transform for a first-order lowpass filter.

$$H_{lp} = \frac{1}{1+\mathrm{j}\omega/\omega_o} \rightarrow H_{bp} = \frac{1}{1+(\mathrm{j}\omega - \mathrm{j}\omega_c)/\omega_o} \tag{6.47}$$

The most efficient implementation technique is the direct synthesis of the transfer function. This was worked out in equation 3.25 for the linear transformation of a first-order lowpass filter. The realization of equation 3.25 is given in fig. 6.45. The required building blocks are summators, amplifiers and integrators.

The realization of translated higher order lowpass filters with only poles in their transfer function can be done by cascading the structure of fig. 6.45. The translation adds a complex term, equal to $\mathrm{j}\omega_c$, to the position of each pole. Complex conjugate complex conjugate poles will not be complex conjugate anymore and all poles will

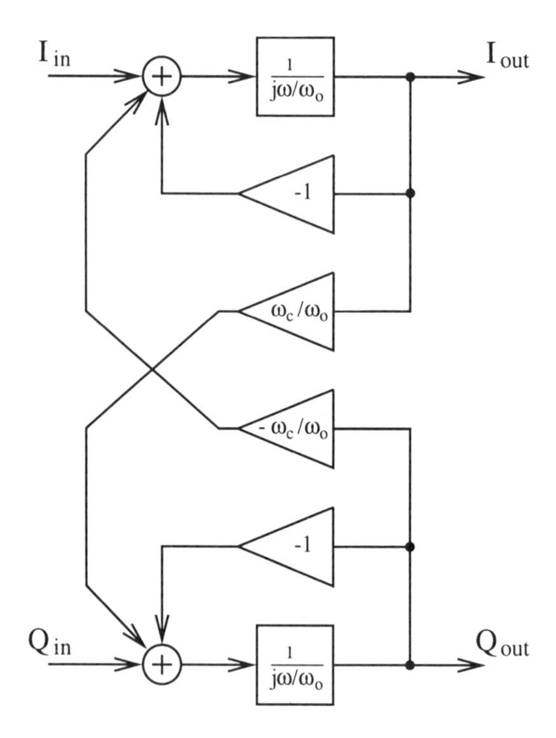

Figure 6.45. Blockdiagram of one section of an active polyphase filter.

have to be realized separately. For the implementation of an nth-order filter $2n$ integrators are required. This is the same as when two lowpass filters are used in a zero-IF and it is far less than what would be required when two bandpass filters are used.

The quality of the mirror suppression is determined by the matching in the polyphase filter. Mismatch between the amplifiers and integrators in the two paths results in a crosstalk from negative to positive frequencies and vice versa. In a perfect polyphase filter the response to a negative frequencies is only this negative frequency. In a slightly mismatched polyphase filter this response is not only this negative frequency. At the same time, there will now also appear a small positive frequency component at the output. This determines the mirror signal suppression. The mirror signal is situated on the negative low-IF, the wanted signal is situated on the positive low-IF. Crosstalk from negative to positive frequencies results in a superposition of a small part of the mirror signal on top of the wanted signal and this superposition can not be corrected anymore.

6.6.2 A 92 dB Dynamic Range Low-IF Filter

A frequency translated 5th order lowpass Butterworth filter has been implemented. The filter is designed for high quality applications (e.g. the GSM system) which makes that it has to comply with very high specifications dynamic range specifications. Both the adjacent channel suppression and the mirror signal suppression are performed in the filter. The wanted signal, separated from its neighbor and mirror signals, is available at the output of the filter. It must have a signal-to-noise ratio SNR of 60 dB, even for the weakest possible signal. In some applications the weakest signal that must be detected can be 30 dB smaller than its neighbors. This means that the filter must be able to cope with input signals with a dynamic range DR of 90 dB. The filter must also have a controllable gain which can vary with at least 30 dB.

The filter must have a high mirror signal suppression. The matching of the components in the filter is therefore very important. Especially for those components which determine the amplification in the passband (see section 3.4.3).

Table 6.7 gives the normalized poles of a 5th order lowpass Butterworth filter, it also gives the poles for a lowpass bandwidth of 110 kHz (equivalent to a bandpass bandwidth of 220 kHz) and it gives the values of the uncompensated complex poles for a translated version on a center frequency of 250 kHz. The filter is designed for a system with a channel spacing of 200 kHz. The five uncompensated complex poles are realized by cascading the structure of fig. 6.45.

Fig. 6.46 gives the active-RC realization of fig. 6.45. Table 6.8 gives the values of the resistors for a standard capacitance of 20 pF. The filter is realized in a standard 1.2 μm CMOS process (described in appendix A). There are several reasons why the filter is realized with the active-RC implementation technique. First of all there is the required matching of the resistors. The matching of high ohmic polyresistors is much

Table 6.7. Pole positions before and after frequency translation for the proposed 5th order Butterworth filter.

POLES	LOWPASS	LOWPASS	BANDPASS
	$BW = 1$ rad/s	$BW = 110$ kHz	$BW = 220$ kHz
p_1	-0.309 +j 0.951 rad/s	-33.9 +j 105 kHz	-33.9 -j 145 kHz
p_2	-0.309 -j 0.951 rad/s	-33.9 -j 105 kHz	-33.9 -j 355 kHz
p_3	-0.809 +j 0.588 rad/s	-89.0 +j 64.7 kHz	-89.0 -j 185 kHz
p_4	-0.809 -j 0.588 rad/s	-89.0 -j 64.7 kHz	-89.0 -j 385 kHz
p_5	-1 rad/s	-110 kHz	-110 -j 250 kHz

Table 6.8. Resistor values for a standard capacitor of 20 pF.

POLES	R	$R/2Q$
p_1	232 kΩ	22.4 kΩ
p_2	232 kΩ	54.8 kΩ
p_3	89.4 kΩ	25.3 kΩ
p_4	89.4 kΩ	42.9 kΩ
p_5	72.4 kΩ	31.8 kΩ

better than the matching of the MOS transistors in OTA-C and MOSFET-C filters. A good matching in these filters requires the use of large $V_{GS} - V_T$'s and with a single 5 V power supply the mismatch can not be made small enough. Here, the mismatch between the high ohmic polyresistor is less than 0.2 %.

A second reason for choosing the active-RC technique is the required dynamic range. After the first stage of the filter the neighbor signals, which can be a lot higher than the wanted signal, have only been partially filtered out. These signals can still be very high. Therefore, the distortion of this first stage must be very small; in this case lower than -80 dB. This can not be realized with a switched-capacitor [Halon JSSC87], OTA-C [Silva KUL92] or MOSFET-C [Plas JSSC91]. It is only with the active-RC

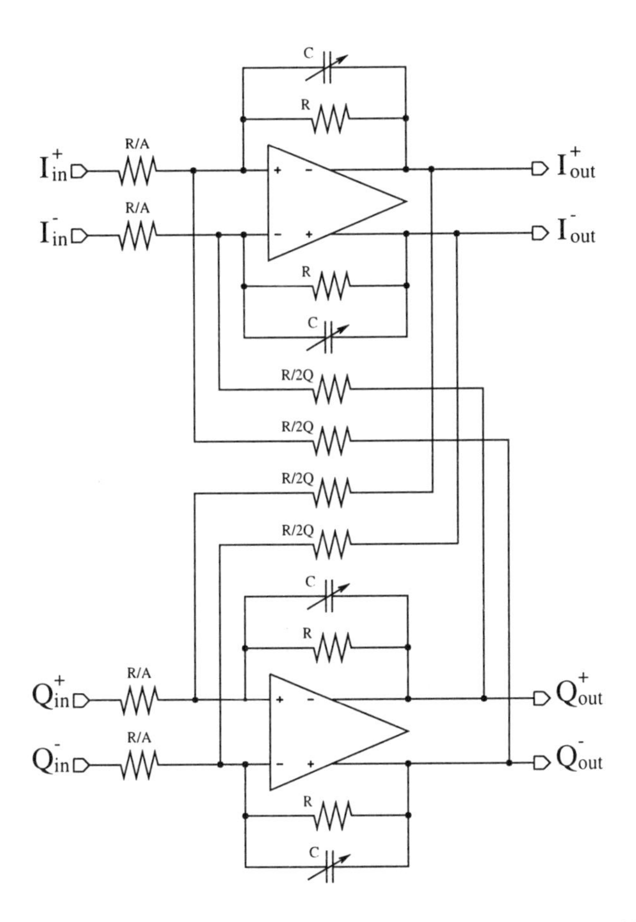

Figure 6.46. The active-RC realization of one section of the polyphase filter.

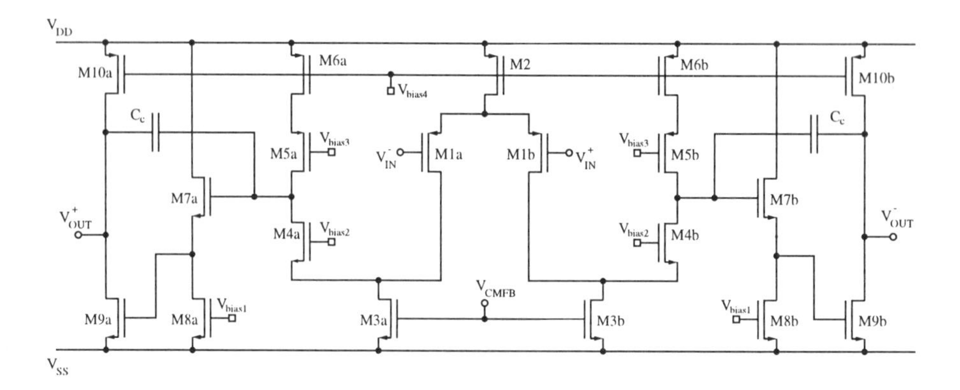

Figure 6.47. The circuit topology of the opamp used in the low frequency active integrated polyphase filter.

implementation technique that such low distortion values can be realized [Durham JSSC, Durh CASII92]. Active-RC is not suited for high frequency applications, but it is ideally suited to realize the required low frequency low-Q filter.

The center frequency of the filter is digitally tunableby means of switchable capacitor banks, as described in [Durh CASII92]. The gain of the first three stages is controllable by means of extra switchable resistors which can be put in parallel with the resistor at the input of the filter stages. The overall gain of the filter can vary from 4 to 256 in steps of 4.

The used opamp topology is given in fig. 6.47. It is a fully differential three-stage Miller-compensated amplifier. The first stage is a folded-cascode topology with pMOS input. With this stage a high open loop gain is achieved. The second stage is a source follower which provides the necessary level-shift to drive the third stage. The third stage is a common source amplifier with low gain, designed to drive relatively low load resistors (20 kΩ). Table 6.9 gives the device sizes for the opamp. The opamp has a gain-bandwidth of 20 MHz, its DC gain is 70 dB and its power consumption is 8.75 mW per opamp operated from a 5 V power supply.

6.6.3 Measurement Results

A microphotograph of the low-IF filter is given in fig. 6.48 [Crols VLSI95]. The chip is realized in a 1.2 μm standard CMOS process. The total chip area is 7.5 mm^2. Half the area is occupied by the switchable capacitor banks. The noise requirements make that the capacitors have to be large. The extra, switchable, capacitors enlarge the total capacitor area with about 50 %. The chip runs on a single power supply voltage of 5 V. The total power consumption is 90 mW.

Table 6.9. Device sizes for the opamp used in the active integrated polyphase filter realization.

DEVICE	TYPE	SIZES
M1	pMOS	$w = 16.8\ \mu m$, $l = 1.5\ \mu m$
M2	pMOS	$w = 121.6\ \mu m$, $l = 1.5\ \mu m$
M3	nMOS	$w = 52.2\ \mu m$, $l = 1.2\ \mu m$
M4	nMOS	$w = 35.2\ \mu m$, $l = 1.2\ \mu m$
M5	pMOS	$w = 121.6\ \mu m$, $l = 1.5\ \mu m$
M6	pMOS	$w = 121.6\ \mu m$, $l = 1.5\ \mu m$
M7	nMOS	$w = 10\ \mu m$, $l = 1.2\ \mu m$
M8	nMOS	$w = 35.2\ \mu m$, $l = 1.2\ \mu m$
M9	nMOS	$w = 69\ \mu m$, $l = 1.2\ \mu m$
M10	pMOS	$w = 240\ \mu m$, $l = 1.5\ \mu m$
Cc	Capacitor	$C = 2$ pF

Figure 6.48. Microphotograph of the low-IF filter, an active integrated polyphase filter.

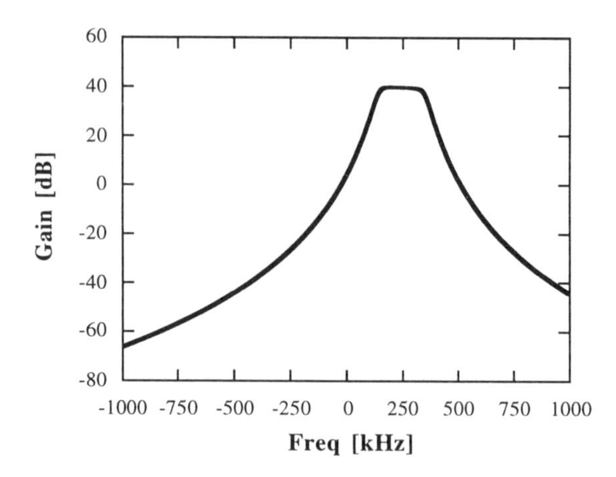

Figure 6.49. The simulated filter transfer function for positive and negative frequencies.

Fig. 6.49 gives the simulated filter transfer function for positive and negative frequencies. Fig. 6.50 shows the measured transfer functions. The full line is the wanted transfer function. The dotted line is the mismatch generated parasitic transfer function from positive to negative frequencies and vice versa. Only crosstalk from negative to positive frequencies is important and fig. 6.50 demonstrates that the mirror signal suppression is more than 64 dB in the complete passband. The main contributor to the frequency crosstalk is the mismatch between the input resistors of the first filter stage (see section 3.4.4). The passband amplitude ripple is less than 0.1 dB.

6.7 CONCLUSION

The possible use of advanced sub-micron and deep sub-micron CMOS technologies for RF circuit design yields many advantages : a lower cost and, in combination with the use of new transceiver architectures, the possibility to realize a true single chip transceiver implementation with all analog and digital signal processing integrated on one die. Problem is however the moderate performance of present day sub-micron CMOS technologies. Studying the possibilities of CMOS RF integration implies therefore that new circuit design techniques have to be developed which exploit the properties of the MOS transistor to realize RF building blocks that achieve performances that are comparable to or better than the performances of today's bipolar and BiCMOS realizations.

In this chapter the implementation of the different types of RF building blocks in sub-micron CMOS has been examined. Design aspects proper to the realization

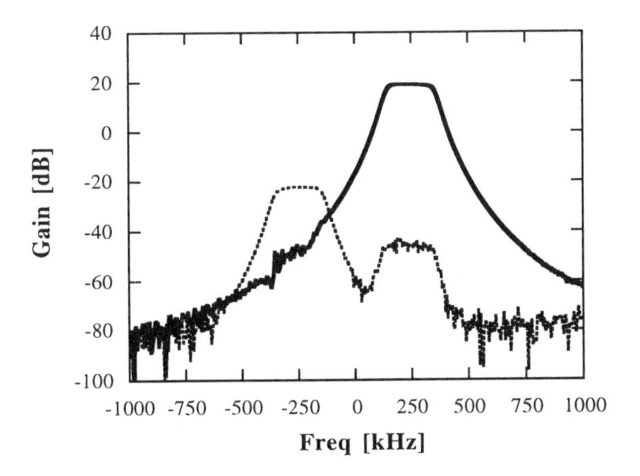

Figure 6.50. The measured transfer functions of the integrated active polyphase filter.

of LNA's, amplifiers, upconversion and downconversion mixers have been discussed. Many chip realization and measurement results were presented and the obtained results prove that by using circuit topologies specially optimized for the use of MOS devices, the realization of a CMOS transceiver front-end becomes possible.

- The high intrinsic linearity of the MOS transistor, when compared with a bipolar transistor, was used to design a highly linear downconversion mixer with a 1.5 GHz input bandwidth. The property that downconversion mixers require only a low frequency output bandwidth was used to realize a circuit topology which requires only a very low power consumption.

- This downconversion mixer has been used as part of a realization of a double quadrature low-IF receiver. The double quadrature receiver is a newly developed topology with which very high quadrature accuracies can be achieved, independent of mismatch and without requiring any tuning or trimming. A phase error of less than 0.3° at 900 MHz has been achieved. This is about 10 times lower than the phase error of classic realizations.

- Upconversion mixers are the most critical building blocks for CMOS implementation because they require an output buffer which has to drive a low impedance off-chip load of 50 Ω at high frequencies. A low impedance can with an nMOS transistor only be realized with a large power consumption and a large transistor, but this would add a lot of parasitic capacitance on the output node. A hand calculation model for planar inductors on lowly doped silicon substrates was introduced which can be used to make a structured and optimized design of an inductor loaded amplifier. In an inductor loaded amplifier the inductor can be used to compensate the parasitic capacitance and in this way the power consumption can be reduced. The accuracy of the modeled inductance is better than 10 % for a wide range of possible inductor layouts.

- A double quadrature upconversion mixer, that uses inductors to compensate for parasitic capacitances, was realized. Again the double quadrature structure was used to achieve a very high quadrature accuracy independent of mismatch and without requiring any tuning or trimming. A sideband suppression of more than 30 dB over a 200 MHz band around 900 MHz has been achieved.

In general, it can be concluded that sub-micron CMOS can be used to realize analog transceiver front-ends for high performance applications in the 900 MHz region, like GSM. Deep sub-micron CMOS will allow for the realization of building blocks up to 2 GHz. In some cases, like for instance the realization of LNA's and linear mixers, deep sub-micron CMOS may even proove to be more suited than silicon bipolar. CMOS is not very well suited for the realization of output buffers loaded by 50 Ω, but the need for such a buffer can be reduced to only one by using new receiver and transmitter topologies which do not require the use of off-chip components and with the evolution to deep sub-micron CMOS good performance will also be possible for this building block.

7 REALIZING A CMOS TRANSCEIVER

7.1 INTRODUCTION

There are many more aspects to the design of an integrated transceiver than just designing its building blocks. The output and input structures of the different building blocks must for instance be designed in such a way that they can drive each other. DC-voltage levels and impedances must be matched. For the layout special attention must be given to a careful floorplanning in order to avoid undesired coupling between sensitive nodes and reduce interconnection lengths.

In this chapter, the aspects important for the realization of a highly integrated transceiver in CMOS are discussed. In chapter 6 the realization of RF CMOS building blocks as stand-alone components was discussed; here, the goal is to discuss some aspects of combining different RF building blocks to obtain a highly integrated transceiver realization. At the end of this chapter a transceiver chip is presented to give an impression of the possible die size and floor plan for a highly integrated transceiver chip. The full study of this transceiver chip realization and all its building block's interactions is considered to be future work. It can be expected that starting from this transceiver's analysis, several design cycles will still be necessary before a true fully functional CMOS RF transceiver can be reported.

7.2 COMBINING BUILDING BLOCKS IN A CMOS TRANSCEIVER

7.2.1 The VCO and Prescaler

In the previous chapters the aspects of the synthesizer design have barely been mentioned. This is because synthesizer design is considered to be a complete field of research on its own. The on-chip integration of the synthesizer is however very important for the realization of a highly integrated CMOS transceiver.

The synthesizer contains two high frequency components : the VCO and the prescaler prescaler. The prescaler is a digital circuit of which its high frequency parts require however an analog design approach in order to achieve the required operating frequency [Cran ESSC95]. It is, in a highly integrated transceiver realization, a fully embedded component (meaning it has no input coming from or output going to the outside world). Its influence on the complete transceiver realization will be through parasitic coupling of its digital switching transients to the other building blocks. The difference with other digital circuitry present, is the high frequencies on which the prescalar's components operate, giving a higher parasitic coupling and an interference on frequencies at which the receive and transmit path are more sensitive.

For the VCO it is important that it is implemented as a true fully integrated realization, meaning a realization without any sensitive high frequency node brought to the outside. It is only in this way that effects like unwanted carrier transmission and detuning caused by coupling between the VCO and the power amplifier can be reduced. A relaxation oscillator can be integrated very well. Its phase noise performance is however too low for practical application in high performance RF implementations. A low phase noise is, especially for the GSM system, one of the main specification for the frequency synthesizer. The use of LC-tank oscillators LC oscillator is therefore required. Fig. 7.1 shows a basic block schematic for the circuit topology of a CMOS voltage controlled oscillator (VCO) [Cran VLSI96]. MOS devices are very well suited for the realization of very low noise VCO's. They are in fact more suited than bipolar transistors [Stey EL94, Cran ISSCC95] (see also section 6.4). On-chip integrated spiral inductors can be used to realize the VCO [Cran VLSI96]. This improves the degree of integration that is achieved. Far more important is however the fact that this eliminates the presence of external high frequency nodes in the VCO. This has not only the advantage of giving a much lower sensitivity to detuning caused by coupling with external high frequency signals. Another advantage is a reduction of the power consumption that can be achieved because of the smaller parasitic capacitances on internal nodes. Today, the high phase noise specifications required by for instance the GSM system can be achieved with the use of on-chip spiral inductors [Cran VLSI96, Cran CICC97].

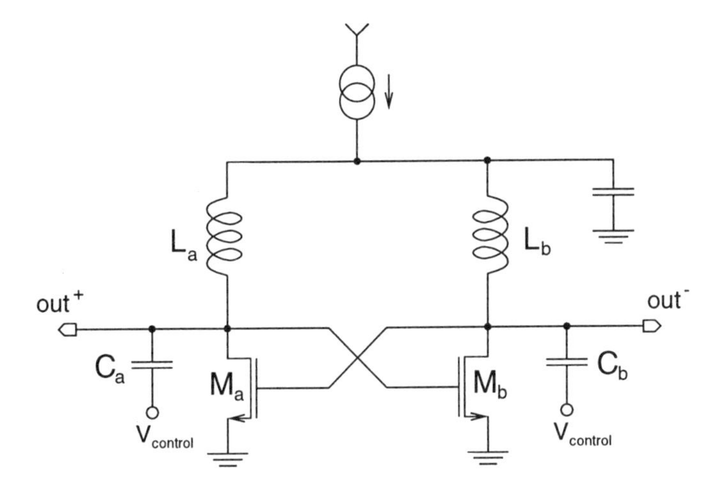

Figure 7.1. A VCO topology that can be fully integrated by using on chip spiral inductors..

7.2.2 The Quadrature Generator

A required quadrature accuracy is 35 dB (or 1° phase error and 150 mdB amplitude error) can be achieved with one two-stage polyphase filter. In this case, there is no need for the use of the highly accurate double quadrature structure as presented in sections 6.5.3 and 6.5.4. This has the advantage that the complexity and power consumption of both the upconverter and downconverter is reduced. Each time only two mixers are required and there is no need for a quadrature generator in the high frequency signal path of the receiver and the transmitter. Only one polyphase filter that serves both the receiver and transmitter is required.

The polyphase filter is placed directly after the VCO and it loads therefore its LC-tank. Because of the partially resistive nature of the impedance of the polyphase filter, direct loading can not be allowed. It is therefore important that when a polyphase filter is used as quadrature generator in a complete analog transceiver front-end realization, special attention is given to the in- and output buffering of the polyphase filter. Fig. 7.2 shows one of the four parallel sections of a two-stage polyphase filter in combination with a possible input and output buffer. For the input buffer a source follower is used. This makes the input impedance of the quadrature generator purely capacitive and its inputs can therefore be coupled directly to the VCO's outputs. Extra DC biasing for the input buffer via capacitive coupling and a large resistor (e.g. 100 kΩ) has to be provided though. Of course, the input capacitance has to be added to the total value of the capacitor for the LC resonance tank of the VCO. The proposed output buffer for the quadrature generator is a common gate amplifier loaded with a resistor. The

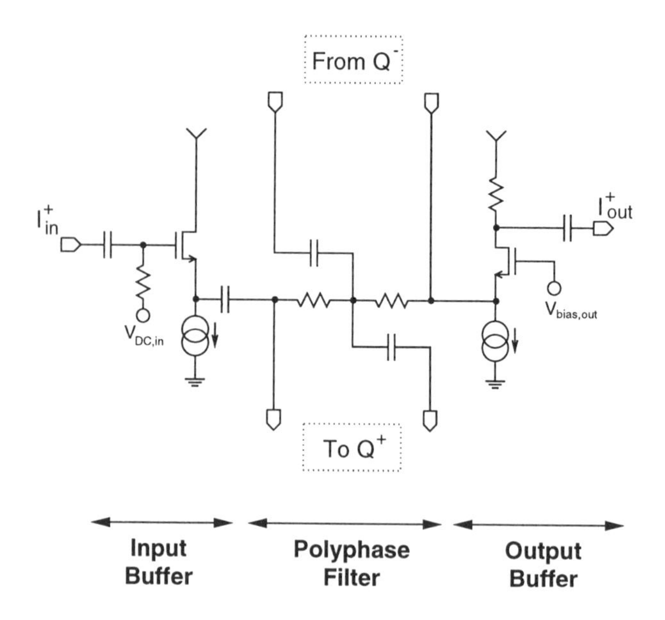

Figure 7.2. One of the four parallel sections of a proposed quadrature generator topology based on a two stage polyphase filter that is provided with input and output buffering.

resistor allows to set the overall gain of the quadrature generator and it also ensures that the quadrature generator is capable of driving both capacitive and resistive loads.

7.2.3 Downconversion

When integrating the downconverter presented in section 6.2.2 as part of a full transceiver front-end, two aspects are important. First of all, it is important to carefully study where a single-ended to differential conversion can be performed in the receive path. Secondly, a circuit topology must be chosen which intrinsically keeps the parasitic self-mixing (RF to RF and LO to LO crosstalk) low.

Fig. 7.3 shows a circuit topology for a quadrature downconverter. It is based on the downconversion mixer presented in section 6.2.2. It does however not use the four-transistor cross-coupled structure which was presented there. Only two MOS transistors operating in the linear region are used per downconverter. This still gives a four quadrant multiplier, but it has the disadvantage that the local oscillator signal is now not only mixed with the RF signal itself but also with the DC biasing of the RF signal. The result of this parasitic mixing operation is however a high frequency signal and this is fully suppressed by the capacitors C_v on the virtual ground nodes.

The advantage of the proposed two-transistor MOS mixing structure is that the downconverter implements in this way also the single-ended to differential conversion for the RF signal. This is an operation which is very difficult to implement at high frequencies. The use of a differential (or balanced) mixer structure is however important, because only in this way a four-quadrant multiplication can be performed. In the proposed downconversion mixer structure this is achieved by realizing the single-ended to differential conversion during the mixing process.

The drawback of the mixer topology proposed in fig. 7.3 is that it does not have a good performance concerning parasitic crosstalk (resulting in self-mixing). By using only a single-ended RF signal, crosstalk of this signal to a source or drain of one of the mixing transistors is not directly cancelled by crosstalk of an equal but opposite (negative) RF signal. This is different from the four transistor cross-coupled downconversion mixer topology presented in section 6.2.2. In that topology crosstalk of the two differential components of the RF signal directly compensate each other and the remaining crosstalk is then proportional to the mismatch between the parasitic capacitance of the mixer transistor, rather than proportional to the parasitic capacitances themselves. In the topology of fig. 7.3 the parasitic crosstalk is a common-mode signal and it is therefore suppressed after downconversion with the common-mode feedback of the fully differential low frequency opamp. At low frequencies matching can be good and a good suppression of the crosstalk components may therefore be suspected. In upconversion mixers, the crosstalk signal is a high frequency common-mode signal. An accurate suppression of common-mode signals at high frequency is much more difficult and a good topology selection is thus even more important for the upconverter.

7.2.4 Upconversion and Preamp

Fig. 7.4 presents a circuit topology for an upconversion mixer. It is again based on MOS transistors operated in the linear region connected to a virtual ground point. The virtual ground is here however not generated by means of a feedback loop over an opamp but by means of a current conveyor (current amplifier, [Touma 1990]) which operates at the same time as the preamp. The current conveyor has to be able to drive a 50 Ω load, to give sufficient current gain and to provide a low enough input impedance at 900 MHz. A current conveyor has the advantage that it can be realized as an open loop structure (for instance a current mirror) and therefore achieve a high operating frequency.

The proposed upconverter topology uses a single-ended current conveyor and all modulated MOS transistors are connected to the same virtual ground node. This has again the advantage that the differential to single-ended conversion is realized during the upconversion process, making the high frequency signal path a single-ended signal path. The drawback is however that the signal component generated by the square law operation of the MOS transistor due to its dependance of g_{DS} on V_{DS} is

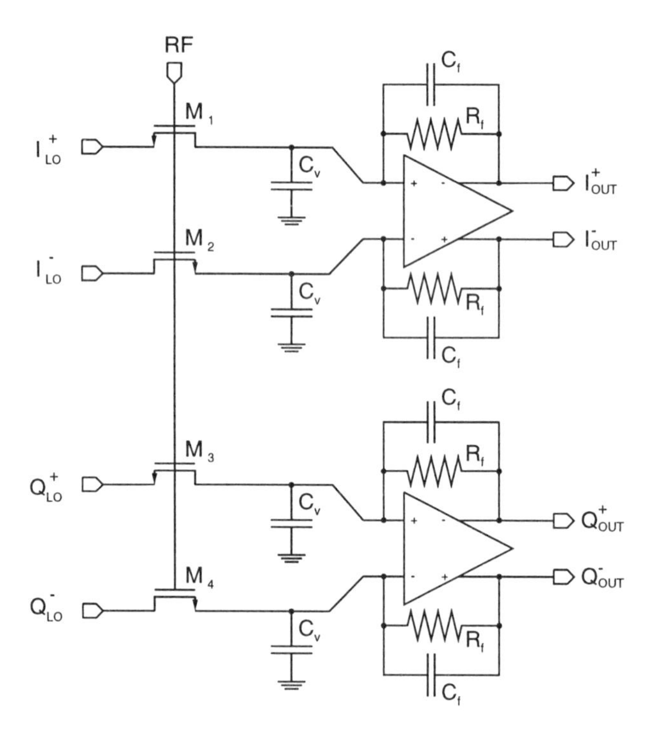

Figure 7.3. A CMOS quadrature downconverter topology that performs a four-quadrant multiplication and a single-ended to differential conversion.

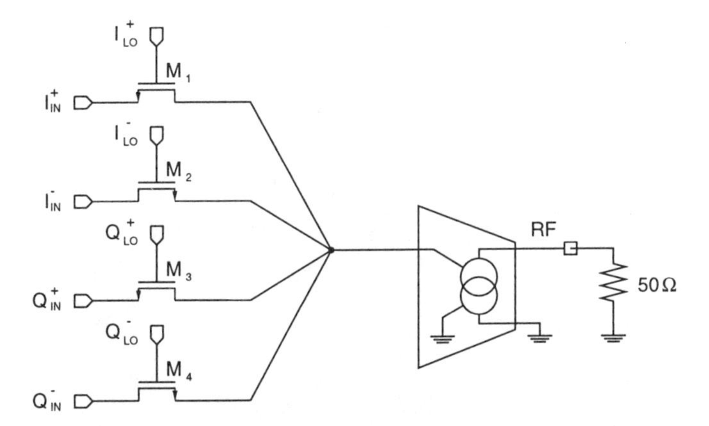

Figure 7.4. The proposed upconverter topology for a highly integrated CMOS transceiver chip realization.

not suppressed anymore by the differential structure and its common-mode suppression [Borre CICC97]. This square law operation is performed on the baseband signals and its result is a low frequency signal. In the current conveyor extra circuitry has to be provided which suppresses this low frequency signal and separates it from the wanted high frequency signal.

It is important to notice that the LO signals are the only high frequency input signals present and that they are provided as fully differential signals. All unwanted parasitic coupling of these signals to the sensitive high frequency source or drain of the mixing transistors is directly combined in one node. They compensate each other thus directly and parasitic coupling is reduced to the effects of mismatches between parasitic capacitances.

7.2.5 Realization of a CMOS Transceiver

Fig. 7.5 shows as an example a microphotograph and the floorplan of a realized highly integrated analog transceiver front-end chip. The chip has been realized in a standard 0.4 μm CMOS process. The chip's die area is 12 mm^2. The transceiver's architecture is the low-IF receiver / direct upconversion architecture which was derived in chapter 5. All the transceiver's different building blocks are designed to comply with all the specifications required for the GSM system as derived in chapter 5. An overview of all these building block specifications can be found in section 5.4.6. More information on the actual circuit realization of the transceiver's different building blocks and their measurement results can be found in [Cran CICC97, Borre CICC97, Jans VLSI97]. The study of the interaction of the different building blocks in the full transceiver is

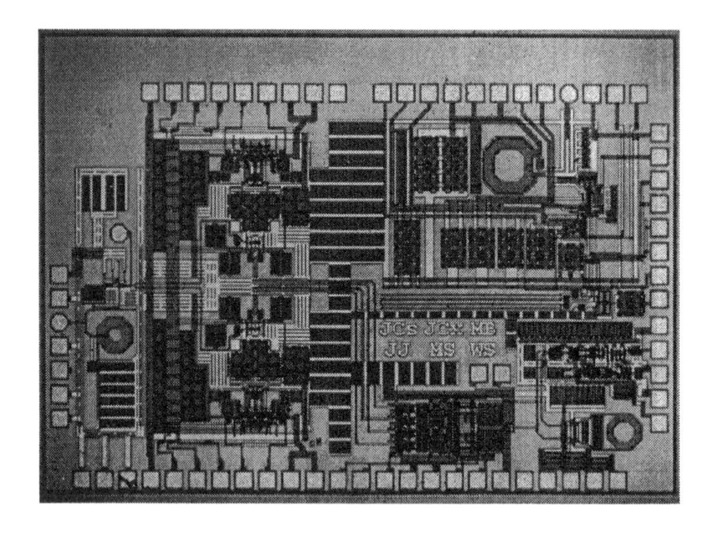

(a)

LNA
Downconverter I
Downconverter Q
Upconverter
Prescaler
Quadrature Generator
VCO

(b)

Figure 7.5. a) Microphotograph of a highly integrated CMOS transceiver chip and b) its floorplan.

considered future work for which this CMOS transceiver implementation can be used as a test vehicle.

7.3 CONCLUSION

In this chapter an indication was given of the many extra aspects which must be considered when a highly integrated CMOS transceiver front-end is realized. This realization is much more than only combining the different building blocks. A first realization of a CMOS transceiver realized in a deep sub-micron process was presented to give the reader an impression of die size and floorplan. Future work will include the full study of this CMOS transeiver front-end realization and a further optimization of its design.

8 GENERAL CONCLUSIONS

When digital signal processing techniques were introduced into the field of wireless communication, they brought many changes. Digital signal processing, in combination with digital data communication, has allowed for the implementation of transparent and highly reliable wireless networks. Wireless networks offer to their users the automatic set-up of point-to-point communication links over a medium (free air) which is originally not suited for this type of application because it has to be shared by all users. With the availability of transparent point-to-point communication, the world of wireless communication has been opened for personal communication applications. Wireless communication is now not only limited anymore to a selected number of professional applications. Today, many wireless services becoming available for people to be used in their daily life. The market of personal communications is an enormously growing mass market and its impact on the evolution of wireless communication equipment has been accordingly.

In order for wireless terminals to be suited for the mass market they need to be priced low and they need to have a small physical size. Large investment into the development of new handheld terminals for personal wireless communications have been made, resulting in a constant drop of their price and weight over the past five years. Today, handheld wireless terminals are priced below 300 US$ and they weigh less than 200 g. The number of wireless applications is however constantly increasing and new generation of wireless interfaces will have to be developed which are priced well below 100 US$ and weigh less than 50 g. These wireless interfaces will become a standard attribute of electronic equipment like laptops and personal digital assistants or even complete wireless multi-media terminals.

The development of these next-generation wireless interfaces will require the evolution from what is today a three or four chip-set solution with several external components to a single-chip solution with only a few external components. This has to be combined with a further reduction of the power consumption of the wireless interface as this has a direct impact on the battery size and cost. The integratability and power consumption of the digital part of a wireless transceiver will further improve with the continued downscaling of CMOS technologies to sub-micron and deep sub-micron gate lengths. The bottle-neck for further advancement is the analog front-end which forms the interface between the antenna and the digital signal processor in a wireless transceiver. An increased level of integration and a further reduction of the power consumption can not be achieved by merly using more advanced technologies. What is required is the development of new transceiver front-end architectures.

The heterodyne receiver and transmitter architecture is for already more than 40 years the most commonly used. It has however one big disadvantage. It uses intermediate frequency stages and each one of these stages requires a bandpass filter that can not be integrated on chip. With the ever increasing number of wireless applications, the frequency band of each new application becomes higher and higher, resulting in the need for more and more intermediate stages and thus more external components; an evolution which is the opposite of the market's requirements. The zero-IF receiver topology and the direct upconversion transmitter topology can be integrated in a much better way. Respectively their sensitivity to parasitic baseband signals and coupling with the high power RF signal has limited their widespread use, although good performances for some applications have been demonstrated. Today, combined IF zero-IF architecture are considered to give the best trade-off between integratability and performance.

In this work a new technique for the design of receiver and transmitter architectures was introduced. It is based on the use of complex signals and operations defined on these complex signal to synthesize multi-path architecture. The zero-IF and direct upconversion topology are in fact examples of such multi-path architectures. There are however many more possibilities and this is exactly what the synthesis technique which was presented in chapter 3 offers. The complex signal technique for multi-path analog signal processing systems offers a method to synthesize and analyze sequences of filters, mixers and amplifiers which do not require external components, like passive bandpass filters, anymore. The use of parallel (or multi) analog signal processing paths eliminates this need. The fact that sequences of several mixing stages can be used eliminates the sensitivity problems typical for zero-IF and direct upconversion topologies. In chapter 3 not only the complex signal technique for transceiver front-ends was introduced, this technique was also used in this chapter to develop several new receiver and transmitter topologies. These topologies all have in common that they do not try to find a good trade-off between integratability and performance, as does the combined IF zero-IF topology. Instead, they all combine the advantage of a good integratability with a high performance and low sensitivity to parasitic signals. The low-IF receiver topology is an example of such a new receiver topology. It uses a very low (a few hundred kHz) intermediate frequency which makes that its IF filter can be integrated on chip, while it avoids the problems of interference by parasitic baseband signals.

Design of the topology for a transceiver front-end for a given application is more than only determining a sequence of filters, mixers and amplifiers which performs the required signal transformations in the frequency domain. The noise and signal level capabilities specified for the given application have to be translated in specifications for the individual building blocks of the choosen transceiver architecture. Today, this is, just like the topology selection, a mainly experience-based process. Each type of analog building block has a large number of specifications that must be determined and only experienced designer have a good feeling for the interdependencies between all parameters. New architectures and building blocks require the full and long design cycle down to transistor level before some insight is gained in the interdepedencies of their parameters. This makes the introduction of new architectures and building blocks of course a very lengthy process, while often only a limited optimization is performed (resulting in a too high power and chip area consumption). In chapter 4 a general applicable high-level design technique for wireless transceivers was presented. It offers a method with which the high-level optimization problem can be formally described and solved, making in this way a computer automated implementation of the high-level design problem possible. The proposed high-level design technique is based on a new technique for the behavioral modelling of the different building blocks for wireless transceivers. These behavioral models not only give a high-level representation of the building block's operation, they are fully described by a unique set of five independently variable parameters and they also give a relationship between these parameters and cost parameters like power and chip area consumption. These models are used in combination with a newly developed analysis technique for wireless transceivers, based on the representation of power spectral densities as rational polynomials. This analysis technique allows for a high-level calculation of the performance of a fully specified transceiver topology without having to go down to the transistor level. In combination with an optimization algorithm, it allows for a fast and memory efficient high-level design. A structured high-level design, that includes the evaluation of many alternative topologies and the full optimization of the chosen topology, is made possible in this way.

The wireless transceiver market demands a very high degree of integration and a low cost. In this respect CMOS technologies seem an attractive alternative for the silicon bipolar, BiCMOS and GaAs technologies which are currently used for the integration of analog transceiver front-ends. This advantage is not so much its lower cost per area as this may not compensate for the inherently lower performance of CMOS technologies. More important is the compatability with the technology of the digital baseband signal processor. Only a full CMOS integration of the RF part will allow for the realization of a true single-chip transceiver that has both all analog and digital signal processing implemented on the same die. In chapter 6 the possibilities of CMOS integration of the different types RF building blocks was studied. In order to be suited for RF integration, a technology has three critical requirements : the availability of low noise devices (for the integration of LNA's and VCO's), a high frequency performance of its devices and the capability to drive low impedances (50 Ω) with a good power efficiency. A MOS transistor is an inherently more linear device than a bipolar transistor and with this taken into account it was demonstrated that nMOS transistors are much more suited than bipolar transistors for the realization of the input stage of an LNA. In fact for the same power consumption and linearity, a MOS input stage will generate less noise. The problem which remains is its frequency performance. The f_T of an nMOS transistor is lower than the f_T of an npn transistor. However, with the evolution to deep sub-micron CMOS technologies this difference becomes smaller and nMOS transistors with f_T's well above 50 GHz become available. It is especially under these circumstances that it becomes important to consider CMOS as a viable technology for RF integration. A point on which CMOS is not good, and which will not improve with technology downscaling, is the power required for the realization of a low impedance driver. On-chip high frequency nodes can use impedance levels of 250 to 400 Ω, but off-chip 50 Ω has to be used in order to cope with the higher parasitic capacitances of external components and interconnections. The use of new transceiver architectures which have only one high-frequency output node already greatly reduces this problem. Furthermore, in chapter 6 it was studied how on-chip spiral inductors can be used to further reduce the power consumption of these output drivers. An analytical model for spiral inductors on lowly doped substrates was developed in order to allow for a structured design and evaluation of analog circuits that use spiral inductors. In chapter 6 many chip realization in standard sub-micron CMOS technologies were presented. Not only classical building blocks, like mixers, were implemented in CMOS, but also several complex operators, newly introduced in chapter 3, were implemented in order to demonstrate their capabilities. Examples are the double quadrature downconverter, the double quadrature upconverter and the complex low IF filter. It was demonstrated that performances equal to or better than bipolar realization can be achieved in sub-micron CMOS. The capabilities of deep sub-micron CMOS technologies as alternative for bipolar, BiCMOS and GaAs RF circuit realization seems therefore undoubtable.

With the RF capabilities of deep sub-micron CMOS technologies when used in combination with new transceiver topologies proven, the final goal of the presented work was a first realization of a highly integrated analog transceiver front-end chip in a deep sub-micron CMOS technology. The GSM system at 900 MHz was chosen as demonstrator. The design was required to fully comply with the specifications as given for a 2 Watt handheld mobile station for GSM. In chapter 5 the high-level design of this chip was performed, based on the new topologies and methods derived in chapters 3 and 4. The designed transceiver architecture requires only four extra components : an antenna, an antenna duplexer with blocking filter, a power amplifier and a digital baseband processing chip. In chapter 7 aspects concerning the integration of this highly integrated transceiver and a first realization was presented. Future work will include a careful study of the interaction between the transceiver's building blocks and an optimization of the circuit topologies of the different building blocks.

Remaining Challenges In the on-going research for the realization of a fully integrated single-chip transceiver in CMOS, two challenges remain. One is the integration of the power amplifier. Today, most power amplifiers are class-C hybrid modules built around one or more discrete GaAs transistors. In some applications the output signal's power is as high as 2 Watt; with an efficieny of 50 %, this means also a 2 Watt power consumption in the power amplifier. Both the present level of integration and the power consumption are much too high. Possible integration will therefore require a significant improvement of the efficiency, which will of course also be beneficiary for the overal power consumption of the transceiver. An improvement of the efficiency means however the use of an amplifier stage different from class-C, for example a class-E amplifier, but such amplifiers suffer from serious non-linearity effects. An evolution to this type of power amplifiers is however inevitable.

A second challenge that remains is the coupling via substrate and substrate coupling package between analog and digital circuitry placed on the same die. This is in fact a general problem common to all mixed-signal chip realizations and a lot of research is being done on this topic today [Su JSSC93]. It is however often overlooked that this coupling is not as disturbing for RF applications as it is for other mixed-signal applications. The specific nature of wireless communication (a very small bandwidth signal situated at a high operating frequency) makes that its analog part is only sensitive to a very small part of the digital switching noise that is coupled in. Problem is that presently no tool is available yet with which the level and spectral distribution of coupled-in digital noise can be predicted. A designer still has to opt therefore for the safest solution, the integration of analog and digital circuitry on a different die, while this may not always be necessary for RF applications.

Appendix A
Process Information

In this appendix SPICE model cards are given for the technologies used in chapter 6. Information on passive components is also provided. For the MOS transistors SPICE level 2 model parameters are given because this model and its extraction process are optimized for analog. More information on the different model parameters and its equations can be found in [Laker 1994]. Special attention must be given to the model parameter LAMBDA in the level 2 model. It is each time given for a transistor with a 3 μm gate length l. For a different gate length l, LAMBDA must be recalculated according to the equations as given in [Laker 1994].

The 1.2 μm CMOS Process

nMOS, SPICE level 2 parameters

```
.MODEL NA NMOS LEVEL=2
+ VTO   =700E-3     NSUB =3.50E16    DELTA=1.8
+ NFS   =1.1E11     TOX  =22.5E-9    UO   =550
+ UCRIT=1.50E5      UEXP =0.15       LD   =1.25E-7
+ XJ    =1E-7       CGSO =1.3E-10    CGDO =1.3E-10
+ RSH   =500        JS   =1E-3       PB   =0.60
+ FC    =0.5        CJ   =1.8E-4     MJ   =0.46
+ CJSW  =2.9E-10    MJSW =0.23       LAMBDA=7.0E-3
+ KF    =1E-28      AF   =1
```

Low V_T pMOS, SPICE level 2 parameters

```
MODEL PLA PMOS LEVEL=2
```

```
+ VTO   =-0.70      NSUB =1.97E16    DELTA=3.76
+ NFS   =4.0E10     TOX  =22.5E-9    UO   =195
+ UCRIT=1.5E5       UEXP =0.23       LD   =0.75E-7
+ XJ    =50E-9      CGSO =1.2E-10    CGDO =1.2E-10
+ RSH   =490        JS   =1E-3       PB   =0.60
+ FC    =0.5        CJ   =5.1E-4     MJ   =0.51
+ CJSW =3.4E-10     MJSW =0.51       LAMBDA=5.0E-3
+ KF    =4E-30      AF   =1
```

Passive devices

- **High ohmic polysilicon resistors (HIPO)** :
  $R_{\Box} = 2$ k$\Omega/\Box$, Matching = 0.5 %

- **Polysilicon to implant capacitors** :
  $C_{ox} = 0.750$ fF/μm^2, $R_{\Box,bottomplate} = 11.5$ $\Omega/\Box$, $C_{bottomplate} \approx 1/3 \cdot C_{ox}$

The 0.7 μm CMOS Process

nMOS, SPICE level 2 parameters

```
.MODEL NA NMOS LEVEL=2
+ VTO   =750E-3     NSUB =7.0E16     DELTA=2.0
+ NFS   =1.2E11     TOX  =17E-9      UO   =470
+ UCRIT=1.08E5      UEXP =0.124      LD   =0.1E-6
+ DELL =0.2E-6      WD   =0.12E-6    DELW =0
+ XJ    =50E-9      CGSO =2.1E-10    CGDO =2.1E-10
+ RSH   =480        JS   =1E-3       PB   =0.65
+ FC    =0.5        CJ   =5.0E-4     MJ   =0.33
+ CJSW =2.8E-10     MJSW =0.214      LAMBDA=8.5E-3
+ KF    =1E-28      AF   =1          NLEV =3
+ GAMMA=0.75        KP   =95E-6      PHI=0.795
+ ACM   = 2         HDIF = 2e-6
```

Low V_T pMOS, SPICE level 2 parameters

```
.MODEL PLA PMOS LEVEL=2
+ VTO   =-0.70      NSUB =3.5E16     DELTA=2.6
+ NFS   =3.7E10     TOX  =17E-9      UO   =173
+ UCRIT=1.3E5       UEXP =0.252      LD   =0.06E-6
```

```
+ DELL =0.15E-6    WD   =0.2E-6     DELW =0
+ XJ   =50E-9      CGSO =1.2E-10    CGDO =1.2E-10
+ RSH  =900        JS   =1E-3       PB   =0.78
+ FC   =0.5        CJ   =6.0E-4     MJ   =0.468
+ CJSW =3.6E-10    MJSW =0.302      LAMBDA=11E-3
+ KF   =4E-30      AF   =1          NLEV =3
+ GAMMA=0.53       KP   =35E-6      PHI  =0.760
+ ACM  = 2         HDIF = 2e-6
```

Passive devices

- **High ohmic polysilicon resistors (HIPO)** :
  $R_{\square} = 2\ \mathrm{k}\Omega/\square$, Matching $= 9.18/(w \cdot l) + 0.0073\ \%^2$

- **Polysilicon to implant capacitors** :
  $C_{ox} = 0.750\ \mathrm{fF}/\mu\mathrm{m}^2$, $R_{\square,bottomplate} = 17\ \Omega/\square$, $C_{bottomplate} \approx 1/3 \cdot C_{ox}$,
  Matching $= 0.11\ \%$

Bibliography

[Abidi ACD96] A.A. Abidi, "A 900 MHz CMOS Spread-Spectrum Wireless Transceiver," in *Analog Circuit Design (W. Sansen, J.H. Huijsing and R.J. van de Plassche, eds.)*, Boston : Kluwer Academic Publishers, pp.101-132, 1996.

[Abidi CICC94] A.A. Abidi, "Radio Frequency Integrated Circuits for Portable Communications," *Proc. CICC*, San Diego, pp.151-158, May 1994.

[Abidi IEEE95] A.A. Abidi, "Low-Power Radio-Frequency IC's for Portable Communications," *Proc. of the IEEE*, Vol. 83, no. 4, pp.544-569, April 1995.

[Abidi ISSCC95] A.A. Abidi, "Direct-Conversion Radio Transceivers for Digital Communications," *Proc. ISSCC*, San Francisco, pp.186-187, Feb. 1995.

[Abidi JSSC95] A.A. Abidi, "Direct-Conversion Radio Receivers for Digital Communications," *IEEE J. of Solid-State Circuits*, Vol. 30, no. 12, pp.1399-1410, Dec. 1995.

[Andre JSSC92] V. Andrews et al., "A Monolithic Digital Chirp Synthesizer Chip with I and Q Channels," *IEEE J. of Solid-State Circuits*, Vol. 27, no. 10, pp.1321-1326, Oct. 1992.

[Baban JSSC85] J.N. Babanezhad and G.C. Temes, "A 20-V Four Quadrant CMOS Analog Multiplier," *IEEE J. of Solid-State Circuits*, Vol. SC-20, no.6, pp.1158-1168, Dec. 1985.

[Balt ACD93] P. Baltus and A. Tombeur, "DECT Zero-IF Receiver Front End," in *Analog Circuit Design (W. Sansen, J.H. Huijsing and R.J. van de Plassche, eds.)*, Boston : Kluwer Academic Publishers, Vol. 2, pp.295-318, 1993.

[Based ESSC94] Ph. Basedau and Q. Huang, "A 1 GHz, 1.5V Monolithic LC Oscillator in 1-μm CMOS," *Proc. ESSCIRC*, Ulm, pp.172-175, Sept. 1994.

[Bast ICMTS95] J. Bastos et al., "Mismatch Characterisation of Small Size MOS Transistors," *Proc. ICMTS*, pp.271-276, March 1995.

[Bird RFD93] J. Bird, "An Introduction to Noise Figure," *RF-design*, pp.78-83, March 1993.

[Borre CICC97] M. Borremans and M. Steyaert, "A 2V, Low Power, Single-Ended 1 GHz CMOS Direct Upconversion Mixer," accepted for publication in *Proc. CICC*, May 1997.

[Bout RFD89] N. Boutin, "Complex Signals," *RF-design*, pp.27-33, Dec. 1989.

[Briant ACD96] F. Brianti et al., "High Integration CMOS RF Transceivers," in *Analog Circuit Design (W. Sansen, J.H. Huijsing and R.J. van de Plassche, eds.)*, Boston : Kluwer Academic Publishers, pp.25-38, 1996.

[Brown ICLMR85] R.A. Brown, R.J. Dewey and C.J. Collier, "An Investigation of the Limitations in a Direct Conversion Radio on FM-Reception," *Int. Conf. on Land Mobile Radio*, pp.157-164, Dec. 1985.

[Burgh IEDM95] J.N. Burghartz et al., "High-Q Inductors in Standard Silicon Interconnect Technology and its Application to an Integrated RF Power Amplifier," *Proc. IEDM*, pp.29.8.1-29.8.3, 1995.

[Chan ESSC93] P.Y. Chan, A. Rofougaran, K.A. Ahmed and A.A. Abidi, "A Highly Linear 1-GHz CMOS Downconversion Mixer," *Proc. ESSCIRC*, Sevilla, pp.210-213, Sept. 1993.

[Chang IEDL93] J.Y.-C. Chang, A.A. Abidi and M. Gaitan, "Large Suspended Inductors on Silicon and Their Use in a 2-μm CMOS RF Amplifier," *IEEE Electron Device Letters*, Vol. 14, no. 5, pp.246-248, May 1993.

[Chien JSSC94] C. Chien, R. Jain, E.G. Cohen and H. Samueli, "A Single-Chip 12.7 Mchips/s Digital IF BPSK Direct Sequence Spread-Spectrum Transceiver in 1.2 μm CMOS," *IEEE J. of Solid-State Circuits*, Vol. 29, no. 12, pp.1614-1623, Dec. 1994.

[Cran CASII95] J. Craninckx and M. Steyaert, "Low-Noise Voltage Controlled Oscillators Using Enhanced LC-tanks," *IEEE Trans. on Circuits and Systems II*, Vol. 42, no.11, Dec. 1995.

[Cran CICC97] J. Craninckx, M. Steyaert and H. Miyakama, "A Fully Integrated Spiral-LC CMOS VCO Set with Prescalar for GSM and DCS-1800 Systems," accepted for publication in *Proc. CICC*, May 1997.

[Cran ESSC95] J. Craninckx and M. Steyaert, "A 1.75GHz/3V Dual-Modulus Divide-by-128/129 Prescaler in 0.7 μm CMOS," *Proc. ESSCIRC*, Lille, pp.254-257, Sept. 1995.

[Cran ISSCC95] J. Craninckx and M. Steyaert, "A CMOS 1.8 GHz Low-Phase-Noise Voltage Controlled Oscillator with Prescaler," *Proc. ISSCC*, San Francisco, pp.266-267, Feb. 1995.

[Cran VLSI96] J. Craninckx and M. Steyaert, "A 1.8 GHz Low-Phase-Noise Spiral-LC CMOS VCO," *Proc. VLSI Circuits Symposium*, Honolulu, pp.30-31, June 1996.

[Crols CASII97] J. Crols and M. Steyaert, "Low-IF Topologies for High Performance Analog Front-Ends of Fully Integrated Receivers," accepted for publication in *IEEE Trans. on Circuits and Systems II*, 1997.

[Crols ESSC94] J. Crols and M. Steyaert, "A Full CMOS 1.5 GHz Highly Linear Broadband Downconversion Mixer," *Proc. ESSCIRC*, Ulm, pp.248-251, Sept. 1994.

[Crols ESSCC96] J. Crols, P. Kinget and M. Steyaert, "A Double Quadrature Topology for High Accuracy Upconversion in CMOS Transmitters," *Proc. ESSCIRC*, Neuchatel, pp.200-203, Sept. 1996.

[Crols ICCAD95] J. Crols, M. Steyaert, S. Donnay and G. Gielen, "A High-Level Design and Optimization Tool for Analog RF Receiver Front-Ends," *Proc. ICCAD*, pp.550-553, Nov. 1995.

[Crols ISSCC95] J. Crols and M. Steyaert, "A Fully Integrated 900 MHz CMOS Double Quadrature Downconverter," *Proc. ISSCC*, San Francisco, pp.136-137, Feb. 1995.

[Crols JSSC95a] J. Crols and M. Steyaert, "A 1.5 GHz Highly Linear CMOS Downconversion Mixer," *IEEE J. of Solid-State Circuits*, Vol. 30, no. 7, pp.736-742, July 1995.

[Crols JSSC95b] J. Crols and M. Steyaert, "A Single-Chip 900 MHz CMOS Receiver Front-End with a High Performance Low-IF Topology," *IEEE J. of Solid-State Circuits*, Vol. 30, no.12, pp.1483-1492, Dec. 1995.

[Crols VLSI95] J. Crols and M. Steyaert, "An Analog Integrated Polyphase Filter for a High Performance Low-IF Receiver," *Proc. VLSI Ciruits Symposium*, Kyoto, pp.87-88, June 1995.

[Crols VLSI96] J. Crols, P. Kinget, J. Craninckx and M. Steyaert, "An Analytical Model of Planar Inductors on Lowly Doped Silicon Substrates for High Frequency Analog Design up to 3 GHz," *Proc. VLSI Circuits Symposium*, Honolulu, pp.28-29, June 1996.

[Czarn CAS86] Z. Czarnul, "Modification of Banu-Tsividis Continuous-Time Integrator Structure," *IEEE Trans. on Circuits and Systems*, Vol. 33, pp.714-716, July 1986.

[Durh CASII92] A.M. Durham, J.B. Hughes and W. Redman-White, "Circuit Architectures for High Linearity Monolithic Continuous-Time Filtering," *IEEE Trans. on Circuits and Systems II*, pp.651-657, Sept. 1992.

[Durham JSSC] A.M. Durham, W. Redman-White and J.B. Hughes, "High-Linearity Continuous-Time Filter in 5-V VLSI CMOS," *IEEE J. of Solid-State Circuits*, Vol. 27, no. 9, pp.1270-1276, Sept. 1992.

[ETSI94] –, "European digital cellular telecommunications system (Phase 1); Radio transmission and reception; DCS extension," *Eurpean Telecommunications Standard Institute*, ETSI 0505-DCS, 1994.

[EW92a] –, "Direct-conversion FM design," *Electronics World*, pp.962-967, Nov. 1992.

[EW92b] –, "Marconi Beginnings," *Electronics World*, pp.74-76, Jan. 1992.

[Faulk EL91] M. Faulkner, T. Mattson and W. Yates, "Automatic Adjustment of Quadrature Modulators," *Electronics Letters*, Vol. 27, no. 3, pp.214-215, Jan. 1991.

[Fenk ACD95] J. Fenk and P. Sehrig, "Low-noise, low-voltage, low-power IF gain controlled amplifiers for wireless communication," in *Analog Circuit Design (W. Sansen, J.H. Huijsing and R.J. van de Plassche, eds.)*, Boston : Kluwer Academic Publishers, Vol. 4, pp.27-44, 1995.

[Freem 1993] E.M. Freeman, *Magnet 5 user guide*, Montreal : Infolytica, 1993.

[Gard 1979] F.M. Gardner, *Phaselock Techniques*, New York : Wiley, 1979.

[Gilb ACD96] B. Gilbert, "Aspects of Translinear Amplifier Design," in *Analog Circuit Design (W. Sansen, J.H. Huijsing and R.J. van de Plassche, eds.)*, Boston : Kluwer Academic Publishers, pp.257-290, 1996.

[Gilb ISSCC83] B.Gilbert, "A Four-Quadrant Analog Divider/Multiplier with .01% Distortion," *Proc. ISSCC*, San Francisco, pp.248-249, Feb. 1983.

[Gilb JSSC68] B. Gilbert, "A Precise Four-Quadrant Multiplier with Subnanosecond Response," *IEEE J. of Solid-State Circuits*, Vol. 3, pp.365-373, Dec. 1968.

[Ging EC73] M.J. Gingell, "Single Sideband Modulation Using Sequence Asymmetric Polyphase Networks," *Electrical Communication*, Vol. 48, pp.21-25, 1973.

[Gray CICC95] P.R. Gray and R.G. Meyer, "Future Directions in Silicon ICs for RF Personal Communications," *Proc. CICC*, pp.83-90, May 1995.

[Green TPHP74] H.M. Greenhouse, "Design of Planar Rectangular Microelectronic Inductors," *IEEE Trans. on Parts, Hybrids and Packaging*, Vol. PHP 10, no. 2, pp.101-109, June 1974.

[Groen CAS91] G. Groenewold, "The Design of High Dynamic Range Continuous-Time Integratable Bandpass Filters," *IEEE Trans. on Circuits and Systems II*, pp.838-852, Aug. 1991.

[Groen JSSC92] G. Groenewold, "A High-Dynamic Range Integrated Continuous-Time Bandpass Filter," *IEEE J. of Solid-State Circuits*, Vol. 27, no. 11, pp.1614-1622, Nov. 1992.

[Grone TCOM76] S.A. Gronemeyer and A.L. McBride, "MSK and Offset QPSK Modulation," *IEEE Trans. on Communications*, Vol. 24, no. 8, pp.809-820, Aug. 1976.

[Haigh 1989] D. Haigh and J. Everard, *GaAs Technology and its Impact on Circuits and Systems*, London : Peter Peregrinus Ltd., 1989.

[Halon JSSC87] K. Halonen, W. Sansen and M. Steyaert, "A Micropower Fourth-Order Elliptical Switched-Capacitor Low-pass Filter," *IEEE J. of Solid-State Circuits*, pp.164-173, April 1987.

[Havi JSSC80] G.L. Haviland and A.A. Tuszynski, "A CORDIC Arithmetic Processor Chip," *IEEE J. of Solid-State Circuits*, Vol. SC-15, no. 1, pp.4-14, Feb. 1980.

[Hull ISSCC96] C.H. Hull, R.R. Chu and J.L. Tham, "A Direct-Conversion Receiver for 900 MHz (ISM Band) Spread-Spectrum Digital Cordless Telephone," *Proc. ISSCC*, San Francisco, pp.344-345, Feb. 1996.

[Jans VLSI97] J. Janssens and M. Steyaert, "A 2.7 V 1 GHz CMOS LNA," accepted for publication in *Proc. VLSI Circuits Symposium*, June 1997.

[Jantz JSSC93] S.A. Jantzi, M. Snelgrove and P.F. Ferguson, "A Fourth-Order Bandpass Sigma-Delta Modulator," *IEEE J. of Solid-State Circuits*, Vol. 28, no.3, pp.282-291, March 1993.

[Karan ISSCC96] A.N. Karanicolas, "A 2.7 V 900 MHz CMOS LNA and Mixer," *Proc. ISSCC*, San Francisco, pp.50-51, Feb. 1996.

[Kaspe PTR83] W.G. Kasperkovitz, "FM Receivers for Mono and Stereo on a Single Chip," *Philips Technical Review*, Vol. 41, no. 6, pp.169-182, June 1983.

[Kim IEDM95] B.K. Kim et al., "Monolithic Planar RF Inductors and Waveguide Structures on Silicon with Performance Comparable to those in GaAs MMIC," *Proc. IEDM*, pp.29.8.1-29.8.4, 1995.

[King CICC96] P. Kinget and M. Steyaert, "A 1 GHz CMOS Upconversion Mixer," *Proc. CICC*, San Diego, pp.10.4.1-10.4.4, May 1996.

[King KUL96] P. Kinget, "Analog VLSI Integration of Parallel Signal Processing Systems," *Ph.D. Thesis*, K.U.Leuven, Leuven, 1996.

[Koul ISSCC93] I. Koullias, J. Havens, I. Post and P. Bronner, "A 900 MHz Transceiver Chip Set for Dual-Mode Cellular Radio Mobile Terminals," *Proc. ISSCC*, San Francisco, pp.140-141, Feb. 1993.

[Kund TCAD86] K.S. Kundert and A. Sangiovanni-Vincentinelli, "Simulation of Nonlinear circuits in the Frequency Domain," *IEEE Trans. on Computer-Aided Design*, Vol. CAD-5, no. 4, pp.521-535, Oct. 1986.

[Laarh 1987] P.J.M. van Laarhoven and E.H.L. Aarts, in *Simulated Annealing : Theory and Applications*, Dordrecht : Kluwer Academic Publishers, 1987.

[Laker 1994] K.R. Laker and W.M.C. Sansen, *Design of Analog Integrated Circuits and Systems*, New York : McGraw-Hill, 1994.

[Lee 1990] E.A. Lee and D.G. Messerschmitt, *Digital Communication*, Boston : Kluwer Academic Publishers, 1990.

[Lee IEEG92] J.-H. Lee and W.-J. Kang, "Designing filters for polyphase filter banks," *IEE Proc.-G*, Vol. 139, pp.363-369, June 1992.

[Manto AICSP93] H.A. Mantooth and P.E. Allen, "A Higher Level Modeling Procedure for Analog Integrated Circuits," *Analog Integrated Circuits and Signal Processing*, Vol. 3, pp.181-195, March 1993.

[Marsh ISSCC95] C. Marshall et al., "A 2.7V GSM Transceiver ICs with On-Chip Filtering," *Proc. ISSCC*, San Francisco, pp.148-149, Feb. 1995.

[Mathw 1992] The Mathworks Inc., *Matlab Refrence Guide*, 1992.

[McDon CICC92] M.D. McDonald, "A DECT Transceiver Chip Set," *Proc. CICC*, pp.10.6.1-10.6.4, May 1992.

[Meyer JSSC74] R.G. Meyer, R. Eschenbach and R. Chin, "A Wide-Band Ultralinear Amplifier from 3 to 300 MHz," *IEEE J. of Solid-State Circuits*, Vol. 9, no. 4, pp.167-175, Dec. 1974.

[Meyer JSSC86] R.G. Meyer, "Intermodulation in High-Frequency Bipolar Transistor Integrated-Circuit Mixers," *IEEE J. of Solid-State Circuits*, Vol. SC-21, no. 4, pp.534-537, Aug. 1986.

[Min CICC94] J. Min, A. Rofougaran, H. Samueli and A.A. Abidi, "An All-CMOS Architecture for a Low-Power Frequency-Hopped 900 MHz Spread Spectrum Transceiver," *Proc. CICC*, pp.379-382, May 1994.

[Muro TCOM81] K. Murota and K. Hirade, "GMSK Modulation for Digital Mobile Radio Telephony," *IEEE Trans. on Communications*, Vol. COM-29, no. 7, pp.1044-1050, July 1981.

[Neuvo ESSC96] Y. Neuvo, "Future Direction in Mobile Communications," *Proc. ESSCIRC*, Neuchatel, pp.35-39, Sept. 1996.

[Nguy JSSC90] N.M. Nguyen and R.G. Meyer, "Si IC-Compatible Inductors and *LC* passive Filters," *IEEE J. of Solid-State Circuits*, Vol. 25, no. 4, pp.1028-1031, Aug. 1990.

[Okan TCE82] T. Okanobu et al., "A Complete Single-Chip AM/FM Radio Integrated Circuit," *IEEE Trans. on Consumer Electronics*, Vol. CE-28, pp.393-408, Aug. 1982.

[Okan TCE92] T. Okanobu, H. Tomiyama and H. Arimoto, "Advanced Low-Voltage Single Chip Radio IC," *IEEE Trans. on Consumer Electronics*, Vol. 38, no. 3, pp.465-475, 1992.

[Optey KUL90] F. Op 't Eynde, "High-Performance Analog Interfaces for Digital Signal Processors," *Ph.D. Thesis*, K.U.Leuven, Leuven, 1990.

[Pelgr JSSC89] M. Pelgrom, A. Duinmaijer and A. Welbers, "Matching Properties of MOS Transistors," *IEEE J. of Solid-State Circuits*, Vol. 24, no. 5, pp.1433-1439, Oct. 1989.

[Pelu ACD96] V. Peluso, M. Steyaert and W. Sansen, "Continuous-Time Bandpass Delta Sigma Modulators in CMOS," in *Analog Circuit Design (W. Sansen, J.H. Huijsing and R.J. van de Plassche, eds.)*, Boston : Kluwer Academic Publishers, pp.171-192, 1996.

[Plas JSSC91] J. Van Der Plas, "MOSFET-C Filter with Low Excess Noise and Accurate Automatic Tuning," *IEEE J. of Solid-State Circuits*, Vol. 26, no. 7, pp.922-929, July 1991.

[Rab ACD93] D. Rabaey and J. Sevenhans, "The challenges for analog circuit design in Mobile Radio VLSI Chips," in *Analog Circuit Design (W. Sansen, J.H. Huijsing and R.J. van de Plassche, eds.)*, New York : Kluwer Academic Publishers, Vol. 2, pp.225-236, 1993.

[Rapal 1994] J. Rapali, "IC Solutions for Mobile Telephones," in *Design of Analog-Digital VLSI Circuits for Telecommunications and Signal Processing (J.E. Franca and Y. Tsividis, eds.)*, London : Prentice Hall, pp.529-568, 1994.

[Razav ICD96] B. Razavi, "Challenges in Portable RF Transceiver Design," *IEEE Circuits and Devices*, Vol. 12, no. 5, pp.12-25, .

[Riley CASII94] T.A.D. Riley and M.A. Copeland, "A Simplified Continuous Phase Modulator Technique," *IEEE Trans. on Circuits and Systems II*, Vol. 41, no. 5, pp.321-328, May 1994.

[Riley JSSC93] T.A.D. Riley, M.A. Copeland and T.A. Kwasniewski, "Delta-Sigma Modulation in Fractional-*N* Frequency Synthesis," *IEEE J. of Solid-State Circuits*, Vol. 28, no. 5, pp.553-559, May 1993.

[Rober 1989] I. Robertson, "Mixer Designs for GaAs Technology," in *GaAs Technology and its Impact on Circuits and Systems (D. Haigh and J. Everard, eds.)*, London : Peter Peregrinus Ltd., pp.281-312, 1989.

[Rofou CICC96] A. Rofougaran et al., "A 900 MHz CMOS Frequency-Hopped Spread-Spectrum RF Transmitter," *Proc. CICC*, San Diego, May 1996.

[Rofou ESSC95] A. Rofougaran et al., "A 1GHz CMOS RF Front-End IC with Wide Dynamic Range," *Proc. ESSCIRC*, Lille, pp.250-253, Sept. 1995.

[Rofou ISSCC96] A. Rofougaran, J. Real, M. Rofougaran and A.A. Abidi, "A 900 MHz CMOS LC-Oscillator with Quadrature Outputs," *Proc. ISSCC*, San Francisco, pp.392-393, Feb. 1996.

[Rofou VLSI94] M. Rofougaran, A. Rofougaran, C. Olgaard and A.A. Abidi, "A 900 MHz CMOS RF Power Amplifier with Programmable Output," *Proc. VLSI Circuits Symposium*, June 1994.

[Salla ISSCC90] D. Sallaerts, D. Rabaey, A. Vanwelsenaers and M. Rahier, "A 270 kbit/s 35mW Modulator IC for GSM Cellular Radio Hand-held Terminals,," *Proc. ISSCC*, San Francisco, pp.34-35, Feb. 1990.

[Seven CICC91] J. Sevenhans, A. Vanwelsenaers, J. Wenin and J. Baro, "An integrated Si bipolar transceiver for a zero IF 900 MHz GSM digital mobile radio front-end of a hand portable phone," *Proc. CICC*, pp.7.7.1-7.7.4, May 1991.

[Seven ISSCC94] J. Sevenhans et al., "An Analog Radio front-end Chip Set for a 1.9 GHz Mobile Radio Telephone Application," *Proc. ISSCC*, San Francisco, pp.44-45, Feb. 1994.

[Shaef VLSI96] D. Shaeffer and T. Lee, "A 1.5 V, 1.5 GHz CMOS Low Noise Amplifier," *Proc. VLSI Circuits Symposium*, Honolulu, pp.32-33, June 1996.

[Shanm 1979] K.S. Shanmugan, *Digital and Analog Communication Systems*, New York : Wiley, 1979.

[Shen ISSCC96] D.H. Shen, C.-M. Hwang, B. Lusignan and B.A. Wooley, "A 900 MHz Integrated Discrete-Time Filtering RF Front-End," *Proc. ISSCC*, San Francisco, pp.54-55, Feb. 1996.

[Sheng ISSCC96] S. Sheng et al., "A Low-Power CMOS Chipset for Spread Spectrum Communications," *Proc. ISSCC*, San Francisco, pp.346-347, Feb. 1996.

[Silva JSSC92] J. Silva-Martinez, M. Steyaert and W. Sansen, "Design Techniques for High-Performance Full-CMOS OTA-RC Continuous-Time Filters," *IEEE J. of Solid-State Circuits*, pp.993-1001, July 1992.

[Silva KUL92] J. Silva-Martinez, "Design Techniques for High Performance CMOS Continuous-Time Filters," *Ph.D. Thesis*, K.U.Leuven, Leuven, 1992.

[Song JSSC86] Bang-Sup Song, "CMOS RF Circuits for Data Communications Applications," *IEEE J. of Solid-State Circuits*, Vol. SC-21, no.2, pp.310-317, April 1986.

[Song JSSC90] H. Song and C. Kim, "A MOS four-quadrant analog multiplier using simple two-input squaring circuits with source followers," *IEEE J. of Solid-State Circuits*, Vol. SC-25, pp.841-848, June 1990.

[Soyeu EL95] M. Soyeur et al., "Multilevel monolithic inductors in silicon technology," *Electronics Letters*, Vol. 31, no. 5, pp.359-360, March 1995.

[Stetz ISSCC95] T. Stetzler, I. Post, J. Havens and M. Koyama, "A 2.7V to 4.5V Single-Chip GSM Transceiver RF Integrated Circuit," *Proc. ISSCC*, San Francisco, pp.150-151, Feb. 1995.

[Stey ACD93] M. Steyaert and W. Sansen, "Opamp Design towards Maximum Gain-Bandwidth," in *Analog Circuit Design (W. Sansen, J.H. Huijsing and R.J. van de Plassche, eds.)*, New York : Kluwer Academic Publishers, pp.63-85, 1993.

[Stey ACD94] M. Steyaert and J. Crols, "Analog integrated polyphase filters," in *Analog Circuit Design (W. Sansen, J.H. Huijsing and R.J. van de Plassche, eds.)*, Boston : Kluwer Academic Publishers, Vol. 3, pp.149-166, 1994.

[Stey ACD96] M. Steyaert et al., "RF CMOS Design, Some Untold Pitfalls," in *Analog Circuit Design (W. Sansen, J.H. Huijsing and R.J. van de Plassche, eds.)*, Boston : Kluwer Academic Publishers, pp.63-88, 1996.

[Stey EL94] M. Steyaert and J. Craninckx, "1.1 GHz Oscillator using Bond-wire Inductance," *Electronics Letters*, Vol. 30, no. 3, pp.244-245, Feb. 1994.

[Stey ESSCC96] M. Steyaert et al., "RF Integrated Circuits in Standard CMOS Technologies," *Proc. ESSCIRC*, Neuchatel, pp.11-18, Sept. 1996.

[Stey JSSC92] M. Steyaert and R. Roovers, "A 1-GHz Single-Chip Quadrature Modulator," *IEEE J. of Solid-State Circuits*, Vol. SC-27, no.8, pp.1194-1196, Aug. 1992.

[Su JSSC93] D.K. Su et al., "Experimental Results and Modeling Techniques for Substrate Noise in Mixed-Signal Integrated Circuits," *IEEE J. of Solid-State Circuits*, Vol. 28, no. 4, pp.420-430, April 1993.

[Taka ISSCC95] Ch. Takahashi et al., "A 1.9 GHz Si Direct Conversion Receiver IC for QPSK Modulation Systems," *Proc. ISSCC*, San Francisco, pp.138-139, Feb. 1995.

[Tan JSSC95] L.K. Tan, E.W. Roth, G.E. Yee and H. Samueli, "An 800-MHz Quadrature Digital Synthesizer with ECL-Compatible Output Drivers in 0.8 μm CMOS," *IEEE J. of Solid-State Circuits*, Vol. 30, no. 12, pp.1463-1473, Dec. 1995.

[Touma 1990] C. Toumazou, F. Lidley and D. Haigh, *Analogue IC Design : The Current-Mode Approach*, London : Peter Peregrinus Ltd., 1990.

[Tsivi CICC93] Y.P. Tsividis, "Integrated Continuous-Time Filter Design," *Proc. CICC*, pp.6.4.1-6.4.7, May 1993.

[Voorm 1993] J.O. Voorman, "Continuous-Time Analog Integrated Filters," in *Integrated continuous-Time Filters (Y.P. Tsividis and J.O. Voorman, eds.)*, New York : IEEE press, pp.27-29, 1993.

[Voorm USP90] J.O. Voorman, "Asymmetric polyphase filter," *US Patent 4,914,408,1990*, 1990.

[Wack ASSP86] G. Wackersreuther, "On two-dimensional polyphase filter banks," *IEEE Trans. Acoust., Speech, Signal Process.*, Vol. ASSP-34, pp.192-199, Feb. 1986.

[Wamba KUL96] P. Wambacq, "Symbolic Analysis of Large and Weakly Analog Nonlinear Integrated Circuits," *Ph.D. Thesis*, K.U.Leuven, Leuven, 1996.

[Wang ISSCC89] Yun-Ti Wang, Fang Lu and A.A. Abidi, "A 12.5 MHz CMOS Continuous Time Bandpass Filter," *Proc. ISSCC*, Vol. 25, no. 6, pp.198-199, Feb. 1989.

[Wang JSSC90] Yun-Ti Wang and A.A. Abidi, "CMOS Active Filter Design at Very High Frequencies," *IEEE J. of Solid-State Circuits*, pp.1562-1572, Dec. 1990.

[Weav IRE56] D.K. Weaver, "A Third Method of Generation and Detection of Single-Sideband Signals," *Proc. of the Institute of Radio Engineers*, Vol. 44, no. 12, pp.1703-1705, Dec. 1956.

[Wenin ESSC94] J. Wenin, "IC's for Digital Cellular Communication," *Proc. ESSCIRC*, Ulm, pp.1-10, Sept. 1994.

[Wils JSSC91] J.F. Wilson et al., "A Single-Chip VHF and UHF Receiver for Radio Paging," *IEEE J. of Solid-State Circuits*, Vol. 26, no. 12, pp.1944-1950, Dec. 1991.

Index